Recent Developments of Electrodeposition Coating

Recent Developments of Electrodeposition Coating

Editor

Andrzej Zieliński

MDPI • Basel • Beijing • Wuhan • Barcelona • Belgrade • Manchester • Tokyo • Cluj • Tianjin

Editor
Andrzej Zieliński
Institute of Machine Technology and Materials,
Faculty of Mechanical Engineering and Shipbuilding,
Gdańsk University of Technology
Poland

Editorial Office
MDPI
St. Alban-Anlage 66
4052 Basel, Switzerland

This is a reprint of articles from the Special Issue published online in the open access journal *Coatings* (ISSN 2079-6412) (available at: https://www.mdpi.com/journal/coatings/special_issues/electrodepos_coat).

For citation purposes, cite each article independently as indicated on the article page online and as indicated below:

LastName, A.A.; LastName, B.B.; LastName, C.C. Article Title. *Journal Name* **Year**, *Volume Number*, Page Range.

ISBN 978-3-0365-0736-1 (Hbk)
ISBN 978-3-0365-0737-8 (PDF)

© 2021 by the authors. Articles in this book are Open Access and distributed under the Creative Commons Attribution (CC BY) license, which allows users to download, copy and build upon published articles, as long as the author and publisher are properly credited, which ensures maximum dissemination and a wider impact of our publications.

The book as a whole is distributed by MDPI under the terms and conditions of the Creative Commons license CC BY-NC-ND.

Contents

About the Editor vii

Andrzej Zieliński
Special Issue: Recent Developments of Electrodeposition Coating
Reprinted from: *Coatings* 2021, *11*, 142, doi:10.3390/coatings11020142 1

Modestas Vainoris, Henrikas Cesiulis and Natalia Tsyntsaru
Metal Foam Electrode as a Cathode for Copper Electrowinning
Reprinted from: *Coatings* 2020, *10*, 822, doi:10.3390/coatings10090822 5

Agnieszka Ossowska and Andrzej Zieliński
The Mechanisms of Degradation of Titanium Dental Implants
Reprinted from: *Coatings* 2020, *10*, 836, doi:10.3390/coatings10090836 21

Agnieszka Ossowska, Andrzej Zieliński, Jean-Marc Olive, Andrzej Wojtowicz and Piotr Szweda
Influence of Two-Stage Anodization on Properties of the Oxide Coatings on the Ti–13Nb–13Zr Alloy
Reprinted from: *Coatings* 2020, *10*, 707, doi:10.3390/coatings10080707 35

Magdalena Jażdżewska and Michał Bartmański
Nanotubular Oxide Layer Formed on Helix Surfaces of Dental Screw Implants
Reprinted from: *Coatings* 2021, *11*, 115, doi:10.3390/coatings11020115 55

Magda Dziaduszewska, Masaya Shimabukuro, Tomasz Seramak, Andrzej Zieliński and Takao Hanawa
Effects of Micro-Arc Oxidation Process Parameters on Characteristics of Calcium-Phosphate Containing Oxide Layers on the Selective Laser Melted Ti13Zr13Nb Alloy
Reprinted from: *Coatings* 2020, *10*, 745, doi:10.3390/coatings10080745 67

Beata Majkowska-Marzec, Dorota Rogala-Wielgus, Michał Bartmański, Bartosz Bartosewicz and Andrzej Zieliński
Comparison of Properties of the Hybrid and Bilayer MWCNTs—Hydroxyapatite Coatings on Ti Alloy
Reprinted from: *Coatings* 2019, *9*, 643, doi:10.3390/coatings9100643 93

Łukasz Pawłowski, Michał Bartmański, Gabriel Strugała, Aleksandra Mielewczyk-Gryń, Magdalena Jażdżewska and Andrzej Zieliński
Electrophoretic Deposition and Characterization of Chitosan/Eudragit E 100 Coatings on Titanium Substrate
Reprinted from: *Coatings* 2020, *10*, 607, doi:10.3390/coatings10070607 107

Rongfa Zhang, Zeyu Zhang, Yuanyuan Zhu, Rongfang Zhao, Shufang Zhang, Xiaoting Shi, Guoqiang Li, Zhiyong Chen and Ying Zhao
Degradation Resistance and In Vitro Cytocompatibility of Iron-Containing Coatings Developed on WE43 Magnesium Alloy by Micro-Arc Oxidation
Reprinted from: *Coatings* 2020, *10*, 1138, doi:10.3390/coatings10111138 125

Andrzej Zieliński and Michał Bartmański
Electrodeposited Biocoatings, Their Properties and Fabrication Technologies: A Review
Reprinted from: *Coatings* 2020, *10*, 782, doi:10.3390/coatings10080782 135

About the Editor

Andrzej Zieliński is a Full Professor at the Gdańsk University of Technology, and former Head of the Materials Science and Engineering Department and Division of Biomaterials. He is an author of about 300 scientific papers and 9 books/chapters, with close to 400 citations, and a Hirsch index of 14. His main interests include the deposition of ceramic, polymer, and composite biocoatings on titanium surfaces; oxidation of 3D laser-assisted printing of titanium; and degradation of materials. He is the supervisor of 18 Ph.D. students.

Editorial

Special Issue: Recent Developments of Electrodeposition Coating

Andrzej Zieliński

Department of Biomaterials Technology, Institute of Machines Technology and Materials, Faculty of Mechanical Engineering and Shipbuilding, Gdańsk University of Technology, Narutowicza 11/12 str., 80-233 Gdańsk, Poland; andrzej.zielinski@pg.edu.pl

Academic Editor: Charafeddine Jama
Received: 24 January 2021; Accepted: 26 January 2021; Published: 28 January 2021

Coatings are one of the forms of surface modifications of several parts produced in many branches of industry and daily life. Coatings may be applied for the protection of carbon steels, aluminum alloys, and even wood and concrete against environmental influence. The hard coatings decrease the wear resistance. However, even though painting is the most popular method of deposition of coatings on buildings, bridges, ships, etc., several functional coatings are being intensively developed. Such examples include the gas-barrier thin films protecting food [1], the coatings against excessive wear of different parts [2,3], textile coatings, e.g., by nanosilver [4], super-hydrophobic coatings [5], and ceramic–metal coatings for protection against erosion [6]. The coatings are often applied in medicine to make the healing time of load-bearing implants with bone faster, to enhance the antibacterial properties, to make steel implants bioinert and easy to remove, and many others. The examples are numerous [7–12].

This Special Issue is aimed at reviewing the newest achievements, particularly in biocoatings. They may be obtained through many techniques, such as direct electrocathodic deposition, pulse electrocathodic deposition, electrophoretic deposition, micro-arc oxidation, chemical and plasma vapor deposition, magnetron sputtering, pulsed laser deposition, electropolymerization, and the sol–gel method, further described in [13–22]. The biocoatings may be made of ceramics, polymeric, metals, or be composite coatings, all so far proposed in the literature.

The electrodeposition coating is among the most plausible techniques, because it makes it possible to design and obtain coatings with different microstructure, thickness, adhesion, and mechanical, physical, chemical, and biological properties. In this Special Issue, composed of nine papers, such examples are shown. One of them shows the deposition of metallic coatings, four papers consider oxidation processes of titanium alloy, three papers are devoted to composite coatings, and the last is review paper on the materials and methods.

Vainoris et al. [23] focused on metallic copper coatings deposited on a flat surface and 3D foams of Cu substrate. The copper deposition occurred much faster on copper foams than on a flat surface, making the metal foams highly suitable for electrowinning. The mechanism of copper deposition was determined, and the capacities of the double electric layer (DL) were calculated. In particular, the DL capacity was much higher and the charge transfer resistance slightly lower for the Cu foam electrodes. As a consequence of this research, the metal foam electrodes were recommended for use in several electrochemical processes.

Ossowska and Zieliński [24] investigated the behavior of new and already used dental implants and the role of oxide layers. In particular, the possible mechanisms of oxide degradation and its influence on titanium corrosion at inflammation states were considered. The extremely low dissolution of rutile, slightly increasing along with pH, was measured. The diffusion of titanium ions through the oxide layer was shown as negligible. The single important mechanism of corrosion was demonstrated as initiated by the oxide layer damage at the defects caused by either the manufacturing process or implantation surgery. Therefore, a stepwise appearance and development of cracks through the oxide layers could be observed and enhance titanium corrosion.

Ossowska et al. [25] focused on the development of sandwich oxide coatings on a titanium base. Two-stage oxidation resulted in the inner solid layer and the outer nanotubular layer of oxides. Such structure of the coating significantly improved mechanical (hardness) and chemical (corrosion resistance) properties. This new technique may be used to substantially improve the surface of titanium load-bearing implants.

Jażdżewska and Bartmański [26] aimed at increasing the corrosion resistance and improving the biocompatibility by oxidation of a model screw dental implant made of the Ti–13Nb–13Zr alloy. The obtained nanotubular layers were of thickness 30–80 nm. The important difference in roughness was noticed between the top of the helix and its bottom. Uneven oxidation of screw model implants resulted in higher corrosion current and less noble corrosion, also known as pitting.

Dziaduszewska et al. [27] studied the micro-arc oxidation in some Ca- and P-containing electrolytes of the selective laser-melted Ti–13Nb–13Zr alloy to obtain ceramic–ceramic composite coatings. The study showed the voltage as the most significant process parameter influencing the coating characteristic. They obtained the coatings with a high Ca:P ratio, hydrophilicity, early-stage bioactivity, Young's modulus, and hardness close to those of bone, and appropriate adhesion of the coating to the titanium surface preventing delamination. Such coatings are especially suitable for dental implants.

Majkowska et al. [28] investigated deposition by the electrophoretic method of ceramic–ceramic coatings composed of hydroxyapatite and carbon nanotubes achieved as bilayers (subsequent deposition) and hybrid coatings (simultaneous deposition). It was shown that the pure multi-wall carbon nanotubes (CNTs) layer showed the best mechanical and biological properties. Both bilayers and hybrid coatings demonstrated insufficient properties attributed to the presence of soft, porous hydroxyapatite and the agglomeration of CNTs.

Pawłowski et al. [29] studied the ceramic–polymer coatings obtained by electrophoretic deposition and composed of chitosan and Eudragit compounds. The best process parameters were estimated. The Young's modulus of coatings was close to that of human cortical bone. The doping of Eudragit significantly reduced the degradation of coatings in artificial saliva at neutral pH, while maintaining high sensitivity to pH changes. The composite coatings showed a slightly lower corrosion resistance compared to the chitosan coating, and comparable hydrophilicity.

Zhang et al. [30] studied metallic (Fe)–ceramic coatings obtained by micro-arc oxidation on Mg alloys. The deposition of such coatings substantially increased the degradation resistance and in vitro cytocompatibility. The developed coatings exhibited potential in clinical applications.

In their paper, Zieliński and Bartmański [31] reviewed the state of the art in electrodeposition coatings. The developments of metallic, ceramic, polymer, and composite electrodeposited coatings were investigated. The direct cathodic electrodeposition, pulse cathodic deposition, electrophoretic deposition, micro-arc oxidation in electrolytes rich in P and Ca ions, electro-spark, and electro-discharge methods were characterized. The most popular were the direct and pulse cathodic electrodeposition, and electrophoretic deposition. The justification of the development of different coatings was an expected increase in bioactivity, mechanical strength, adhesion of coatings, and antibacterial properties.

Conflicts of Interest: The author declares no conflict of interest.

References

1. Nakaya, M.; Uedono, A.; Hotta, A. Recent progress in gas barrier thin film coatings on PET bottles in food and beverage applications. *Coatings* **2015**, *5*, 987–1001. [CrossRef]
2. Aranke, O.; Algenaid, W.; Awe, S.; Joshi, S. Coatings for automotive gray cast iron brake discs: A review. *Coatings* **2019**, *9*, 552. [CrossRef]
3. Ma, L.; Eom, K.; Geringer, J.; Jun, T.S.; Kim, K. Literature review on fretting wear and contact mechanics of tribological coatings. *Coatings* **2019**, *9*, 501. [CrossRef]

4. Verbič, A.; Gorjanc, M.; Simončič, B. Zinc oxide for functional textile coatings: Recent advances. *Coatings* **2019**, *9*, 550. [CrossRef]
5. Bir, F.; Khireddine, H.; Touati, A.; Sidane, D.; Yala, S.; Oudadesse, H. Electrochemical depositions of fluorohydroxyapatite doped by Cu^{2+}, Zn^{2+}, Ag^+ on stainless steel substrates. *Appl. Surf. Sci.* **2012**, *258*, 7021–7030. [CrossRef]
6. Tiwari, A.; Seman, S.; Singh, G.; Jayaganthan, R. Nanocrystalline cermet coatings for erosion-corrosion protection. *Coatings* **2019**, *9*, 400. [CrossRef]
7. Mandracci, P.; Mussano, F.; Rivolo, P.; Carossa, S. Surface treatments and functional coatings for biocompatibility improvement and bacterial adhesion reduction in dental implantology. *Coatings* **2016**, *6*, 7. [CrossRef]
8. Hou, N.Y.; Perinpanayagam, H.; Mozumder, M.S.; Zhu, J. Novel development of biocompatible coatings for bone implants. *Coatings* **2015**, *5*, 737–757. [CrossRef]
9. Graziani, G.; Boi, M.; Bianchi, M. A review on ionic substitutions in hydroxyapatite thin films: Towards complete biomimetism. *Coatings* **2018**, *8*, 269. [CrossRef]
10. Cometa, S.; Bonifacio, M.A.; Mattioli-Belmonte, M.; Sabbatini, L.; De Giglio, E. Electrochemical strategies for titanium implant polymeric coatings: The why and how. *Coatings* **2019**, *9*, 268. [CrossRef]
11. Sartori, M.; Maglio, M.; Tschon, M.; Aldini, N.N.; Visani, A.; Fini, M. Functionalization of ceramic coatings for enhancing integration in osteoporotic bone: A systematic review. *Coatings* **2019**, *9*, 312. [CrossRef]
12. Duta, L.; Popescu, A.C. Current status on pulsed laser deposition of coatings from animal-origin calcium phosphate sources. *Coatings* **2019**, *9*, 335. [CrossRef]
13. Paital, S.R.; Dahotre, N.B. Calcium phosphate coatings for bio-implant applications: Materials, performance factors, and methodologies. *Mater. Sci. Eng. R Rep.* **2009**, *66*, 1–70. [CrossRef]
14. Guslitzer-Okner, R.; Mandler, D. Electrochemical coating of medical implants. In *Applications of Electrochemistry and Nanotechnology in Biology and Medicine I*; Eliaz, N., Ed.; Springer Science & Business Media: Berlin, Germany, 2011; pp. 291–342.
15. Kulkarni, M.; Mazare, A.; Schmuki, P.; Iglič, A. Biomaterial surface modification of titanium and titanium alloys for medical applications. In *Nanomedicine*; Seifalian, A., de Mel, A., Kalaskar, D.M., Eds.; One Central Press Altrincham: Cheshire, UK, 2014; pp. 111–136.
16. Asri, R.I.M.; Harun, W.S.W.; Hassan, M.A.; Ghani, S.A.C.; Buyong, Z. A review of hydroxyapatite-based coating techniques: Sol-gel and electrochemical depositions on biocompatible metals. *J. Mech. Behav. Biomed. Mater.* **2016**, *57*, 95–108. [CrossRef]
17. Dorozhkin, S.V. Calcium orthophosphates (CaPO4): Occurrence and properties. *Prog. Biomater.* **2016**, *5*, 9–70. [CrossRef]
18. Adeleke, S.A.; Bushroa, A.R.; Sopyan, I. Recent development of calcium phosphate-based coatings on titanium alloy implants. *Surf. Eng. Appl. Electrochem.* **2017**, *53*, 419–433. [CrossRef]
19. Liu, W.; Liu, S.; Wang, L. Surface modification of biomedical titanium alloy: Micromorphology, microstructure evolution and biomedical applications. *Coatings* **2019**, *9*, 249. [CrossRef]
20. Su, Y.; Cockerill, I.; Zheng, Y.; Tang, L.; Qin, Y.X.; Zhu, D. Biofunctionalization of metallic implants by calcium phosphate coatings. *Bioact. Mater.* **2019**, *4*, 196–206. [CrossRef]
21. Yang, J.; Cui, F.; Lee, I.S. Surface Modifications of magnesium alloys for biomedical applications. *Ann. Biomed. Eng.* **2011**, *39*, 1857–1871. [CrossRef]
22. Wan, P.; Tan, L.; Yang, K. Surface modification on biodegradable magnesium alloys as orthopedic implant materials to improve the bio-adaptability: A review. *J. Mater. Sci. Technol.* **2016**, *32*, 827–834. [CrossRef]
23. Vainoris, M.; Cesiulis, H.; Tsyntsaru, N. Metal Foam Electrode as a Cathode for Copper Electrowinning. *Coatings* **2020**, *10*, 822. [CrossRef]
24. Ossowska, A.; Zielinski, A. The Mechanisms of Degradation of Titanium Dental Implants. *Coatings* **2020**, *10*, 836. [CrossRef]
25. Ossowska, A.; Zieliński, A.; Olive, J.-M.; Wojtowicz, A.; Szweda, P. Influence of Two-Stage Anodization on Properties of the Oxide Coatings on the Ti–13Nb–13Zr Alloy. *Coatings* **2020**, *10*, 707. [CrossRef]
26. Jażdżewska, M.; Bartmański, M. Nanotubular OxideLayer Formed on Helix Surfaces of Dental Screw Implants. *Coatings* **2021**, *11*, 115. [CrossRef]

27. Dziaduszewska, M.; Shimabukuro, M.; Seramak, T.; Zieliński, A.; Hanawa, T. Effects of Micro-Arc Oxidation Process Parameters on Characteristics of Calcium-Phosphate Containing Oxide Layers on the Selective Laser Melted Ti13Zr13Nb Alloy. *Coatings* **2020**, *10*, 745. [CrossRef]
28. Majkowska-Marzec, B.; Rogala-Wielgus, D.; Bartmański, M.; Bartosewicz, B.; Zieliński, A. Comparison of Properties of the Hybrid and Bilayer MWCNTs—Hydroxyapatite Coatings on Ti Alloy. *Coatings* **2019**, *9*, 643. [CrossRef]
29. Pawłowski, Ł.; Bartmański, M.; Strugała, G.; Mielewczyk-Gryń, A.; Jażdżewska, M.; Zieliński, A. Electrophoretic Deposition and Characterization of Chitosan/Eudragit E 100 Coatings on Titanium Substrate. *Coatings* **2020**, *10*, 607. [CrossRef]
30. Zhang, R.; Zhang, Z.; Zhu, Y.; Zhao, R.; Zhang, S.; Shi, X.; Li, G.; Chen, Z.; Zhao, Y. Degradation Resistance and In Vitro Cytocompatibility of Iron-Containing Coatings Developed on WE43 Magnesium Alloy by Micro-Arc Oxidation. *Coatings* **2020**, *10*, 1138. [CrossRef]
31. Zieliński, A.; Bartmański, M. Electrodeposited Biocoatings, Their Properties and Fabrication Technologies: A Review. *Coatings* **2020**, *10*, 782. [CrossRef]

Publisher's Note: MDPI stays neutral with regard to jurisdictional claims in published maps and institutional affiliations.

© 2021 by the author. Licensee MDPI, Basel, Switzerland. This article is an open access article distributed under the terms and conditions of the Creative Commons Attribution (CC BY) license (http://creativecommons.org/licenses/by/4.0/).

Article

Metal Foam Electrode as a Cathode for Copper Electrowinning

Modestas Vainoris [1], Henrikas Cesiulis [2,*] and Natalia Tsyntsaru [1]

1. Department of Physical Chemistry, Vilnius University, LT-01513 Vilnius, Lithuania; m.vainoris@gmail.com (M.V.); ashra_nt@yahoo.com (N.T.)
2. JSC Elektronikos Perdirbimo Technologijos, LT-06140 Vilnius, Lithuania
* Correspondence: henrikas.cesiulis@chf.vu.lt; Tel.: +370-(5)-2193178

Received: 31 July 2020; Accepted: 22 August 2020; Published: 25 August 2020

Abstract: The geometry of porous materials is complex, and the determination of the true surface area is important because it affects current density, how certain reactions will progress, their rates, etc. In this work, we have investigated the dependence of the electrochemical deposition of copper coatings on the geometry of the copper substrate (flat plates or 3D foams). Chronoamperometric measurements show that copper deposition occurs 3 times faster on copper foams than on a flat electrode with the same geometric area in the same potential range, making metal foams great electrodes for electrowinning. Using electrochemical impedance spectroscopy (EIS), the mechanism of copper deposition was determined at various concentrations and potentials, and the capacities of the double electric layer (DL) for both types of electrodes were calculated. The DL capacity on the foam electrodes is up to 14 times higher than that on the plates. From EIS data, it was determined that the charge transfer resistance on the Cu foam electrode is 1.5–1.7 times lower than that on the Cu plate electrode. Therefore, metal foam electrodes are great candidates to be used for processes that are controlled by activation polarization or by the adsorption of intermediate compounds (heterogeneous catalysis) and processes occurring on the entire surface of the electrode.

Keywords: metal foam; surface area; electrowinning; Cu electrodeposition; EIS; double electric layer capacitance

1. Introduction

The ever-increasing need for electronics, especially, handheld and portable electronics, and the need to reduce their size and increase their efficiency, generates a lot of various electronics waste all over the globe [1–3]. There are many ways to reclaim used metals in electronic waste; however, electrowinning is a very efficient and quite selective process allowing the recovery of high amounts of various pure metals [4–6]. Metallic foams and porous electrodes have an outstanding potential to be used as a cathode to collect deposited metals because of the functionality of their combined material properties resulting from their specific morphology. There already is great interest in the synthesis of various porous materials such as metal foams, nanowires, porous coatings, thin porous films, etc. [7–15]. Depending on the materials, type of pores (open or closed cells), the porosity and size of pores, such materials have broad application capabilities, from simple ones such as heat transfer or electrodes to more complicated cases of various redox reactions, catalysis, sensing, supercapacitors, or even gas storage because of the high surface area and low density available [9,16–29].

Any solid metal surface that acts as a substrate for electrochemical reactions possesses a certain roughness that can affect in different ways the values of the limiting diffusion current and the exchange current density. On the other hand, if the surface coarseness is relatively small, the limiting diffusion current density does not depend on the surface roughness, and it can be only correlated to the apparent surface of the electrodes. If the surface roughness of electrodes increases, the effective values of the exchange current density are also increased for the process under consideration, which is standardized to the apparent electrode surface area. At the same time, the limiting diffusion current density depends on the surface coarseness due to the decrease of the effective value of the diffusion layer thickness. If the level of the electrode surface coarseness remains low, the change of the limiting diffusion current density can be neglected [30]. In addition, it has been shown that when the metal deposition is controlled by diffusion (particularly silver), the surface with the highest surface roughness had a lower number of active sites but higher deposition efficiency and a higher efficiency of charge transfer [31]. The dependence between surface roughness and deposition efficiency is non-linear; the surface roughness needs to be quite high to affect the deposition efficiency [31,32]. It was proven that when the deposition reaction is controlled by the diffusion, the geometry of the electrode has no significant influence on the reaction [33]. Using very porous or surfaces with high roughness, one can eliminate activation and diffusion overpotentials, making the reaction process controlled by Ohmic effects and thus making the reaction much faster [30,31]. All these effects make porous metal electrodes with pore diameters higher than 50 µm high-performance cathodes for deposition reactions under diffusion control.

The estimation of the active surface area of highly porous conducting materials is also very important. Thus, various in situ or ex situ techniques can be used for these purposes. In situ techniques are preferred, since drying the sample can cause changes in the surface area and/or oxidation of the surface, changing its characteristics. Depending on the material and its porosity, one can use techniques for double electric layer estimations (cyclic voltammetry, initial charge-up dependencies, electrochemical impedance spectroscopy (EIS)), or adapt various adsorption/redox reactions that occur on the surface (underpotential depositions, adsorption measurements, reduction of various dyes, etc.) [29,34–39]. The classical techniques for surface area estimation—liquid permeability, gas adsorption (Brunauer–Emmett–Teller technique)—in some cases can also be used [34,35,39]. However, these techniques require higher amounts of materials and can have quite large error margins, depending on the geometry of the pores and the sample itself. For porous materials that are quite level, and with ordered pores, more sophisticated techniques could be used for porosity estimation such as atomic force microscopy (AFM) or spectroscopic ellipsometry; the latter requires a rather complex modeling [40–42].

EIS is a very powerful and versatile in situ technique that allows not only estimating the true surface area of conducting materials but also investigating the surface and the processes happening at the surface [9,20,23,27,29]. Using the EIS technique, one can investigate both Faradaic and non-Faradaic processes on the surface simultaneously [27,29,34,35,43–47]. Even the size and distribution of the pores can be characterized by employing the EIS technique [45,46]. However, the surface area determined by EIS or any other electrochemical method is not the true surface area, but rather the electrochemically active surface area, which can be much more useful when trying to determine the activity of porous materials for a hydrogen evolution reaction (HER) or other electrochemical reaction [39–47].

In this work, we investigated the deposition of copper on the plate (2D) and foam (3D) copper substrates using voltammetry and EIS. The comparison of 2D and 3D electrodes has been carried out to determine differences in double electric layer formation, charge transfer, diffusion, and deposition rates. These results are important for trying to enhance the potential application of foam electrodes in industry, and particularly for the electrowinning of copper from electronics waste.

2. Materials and Methods

2.1. Materials and Sample Preparation

All of the chemicals used for analysis were of analytical grade (Carl Roth, Karlsruhe, Germany). Solutions have been prepared using deionized water (DI). Solution compositions used for electrochemical experiments are shown in Table 1. The pH of solutions was adjusted using sulfuric acid and controlled by a benchtop pH-meter ProLine Plus (Prosence B.V., Oosterhout, The Netherlands).

Cu plates and Cu foam electrodes served as working electrodes. The Cu foam sheets used to fabricate electrodes were purchased from Alfa Aesar. To characterize commercially available copper foams, we have done some experiments trying to determine the basic characteristics of this foam. Foam density has been determined as gravimetrically being equal to 0.748 g/cm^3, making the porosity of the foam to be around 90.5%. The copper foam has a 3D interconnected porous structure, which can be observed in SEM images (Figure 1). The pore size varies from 1 to 0.1 mm. The surface of the foam is very uneven, making the true surface area of the already porous copper foam even larger.

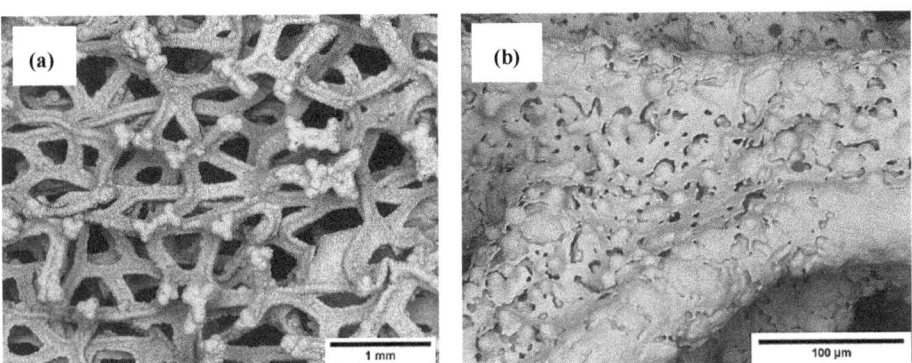

Figure 1. SEM images at low (**a**) and high (**b**) magnification of 3D copper foam.

Working electrodes (copper plates and copper foams) have been washed and degreased using acetone, ethanol, and water in succession and in combination with ultrasonic bath. Both flat and porous samples were 1 cm × 1 cm in geometrical size with both sides conducting. To ensure that the working surface was that of the desired size, other parts of the samples were isolated using insulating plastic spray (PRF 202, Taerosol Oy, Kangasala, Finland). Just before measurements, the native copper oxide layer has been removed by dipping copper samples into 2 M H_2SO_4 solution for 2 s and afterward rinsing with DI water.

Table 1. Composition of solutions used for electrochemical measurements.

$c(CuSO_4)$, M	$c(Na_2SO_4)$, M	pH
0.01	0.49	3.6
0.05	0.45	3.6
0.1	0.4	3.7
0.2	0.3	4.1

2.2. Instrumentation and Methodology

Morphology: The morphology of copper foams has been investigated using a scanning electron microscope (SEM, Hitachi's Tabletop Microscope TM-3000, Tokyo, Japan).

Electrochemical Measurements: Electrochemical measurements (voltammetry, EIS, chronoamperometry, etc.) have been performed using programmable potentiostat/galvanostat AUTOLAB PGSTAT 302N (Metrohm, Utrecht, The Netherlands). The software used for controlling the hardware was Nova 1.11.2.

Conditions of Electrochemical Measurements: A three-electrode system was used for all the electrochemical experiments, where Cu plates or Cu foams were used as working electrodes, circular platinized titanium mesh (Alfa Aesar, Ward Hill, MA, USA) was used as a counter electrode, and Ag/AgCl filled with saturated KCl solution (Sigma-Aldrich, St. Louis, MO, USA) was used as a reference electrode. The distance between the counter and working electrode was fixed at 2.5 cm. All electrochemical experiments have been performed at room temperature. Voltammetry measurements were done using the potential sweep voltammetry technique on Cu plates and Cu foams as working electrodes, starting at open circuit potential and going up to −1.2 V versus Ag/AgCl at a 2 mV/s scan rate. Voltammetry measurements have been performed using all the solutions shown in Table 1. Chronoamperometry experiments were performed at 4 distinct potentials (−0.1, −0.2, −0.4 and −0.6 V versus Ag/AgCl) using different substrates as working electrodes (Cu plates or foams) in 0.1 M $CuSO_4$ and 0.4 M Na_2SO_4 solution. The same amount of electric charge was used to deposit coatings, i.e., 30 C. The current efficiency was calculated using chronoamperometry data and change in substrate mass after deposition.

Electrochemical Impedance Spectroscopy (EIS): Electrochemical impedance spectroscopy (EIS) measurements have been done using a standard three-electrode system, carried out in a frequency range of 10 kHz to 0.1 Hz, using perturbation amplitude of 10 mV. Obtained data were fitted to the equivalent electric circuit model (EEC) using ZView 2.8d software.

3. Results and Discussion

3.1. Copper Foam Characterization

In order to determine how the behavior of copper foams differs from flat surfaces in solutions, voltammetry experiments with different copper sulfate concentrations were carried out; the compositions of the solutions are shown in Table 1. The concentration of the sulfate anion was kept at 0.5 M to maintain the same buffering power in all of the solutions. The obtained polarization curves for the plate and foam electrode are shown in Figure 2, where the ordinate axis is displayed in a logarithmic scale because of a big difference in the current values between tested concentrations. To estimate the influence of porosity on the copper deposition, the geometrical sample size was the same for both Cu plates and Cu foams (1 cm × 1 cm). As can be seen from Figure 2, Cu deposition starts somewhere around −0.075 V versus Ag/AgCl and did not depend on the substrate used. After the peak representing the Cu^{2+} reduction to Cu^0, the current on both surfaces and all the concentrations turns into an almost constant one. The reason for this could be the mass transport limitations because the leveling off of the current depends on the concentration of Cu(II) in the solution. This is also supported by the slight increase of the current with the rise of polarization at higher concentrations (50 mM to 0.2 M), showing that with higher potential, the positive ions are attracted from further away, and the deposition rate increases.

In addition, voltammetry tests also showed that independently of the substrate used, the hydrogen evolution reaction (HER) started in the range of −1.0 to −1.1 V versus Ag/AgCl in the solutions containing 10 and 50 mM of $CuSO_4$. This fact could be attributed to the governing role of pH change in the pre-electrode layer during electrodeposition, and this change seems to be similar for both solutions. However, in the solution containing 0.2 M $CuSO_4$, the HER started around −0.75 V versus Ag/AgCl on both surfaces. It can be linked to the higher rate of copper electrodeposition, and in turn, the pH decrease near the working electrode. Thus, the major difference between the two surfaces can be noted from voltammetry experiments: there was an approximately 3 times higher current on the foam substrate at all potentials in comparison to the flat surface. This difference can be explained by the better hydrodynamic conditions of copper foams substrate: the porous surface allows for faster mass transport and exchange.

For further investigation, the solution containing a similar amount of Cu(II) as in solutions used for the metals recovery from the electronic waste was chosen. Regarding the influence of the surface type on the Cu electrochemical deposition, chronoamperometric measurements have been done in 0.1 M $CuSO_4$ and 0.4 M Na_2SO_4 solution at four fixed potentials: −0.1, −0.2, −0.4, and −0.6 V versus Ag/AgCl, and at a fixed amount of charge passed through the cell (30 C). The results have been summarized and are shown in Table 2.

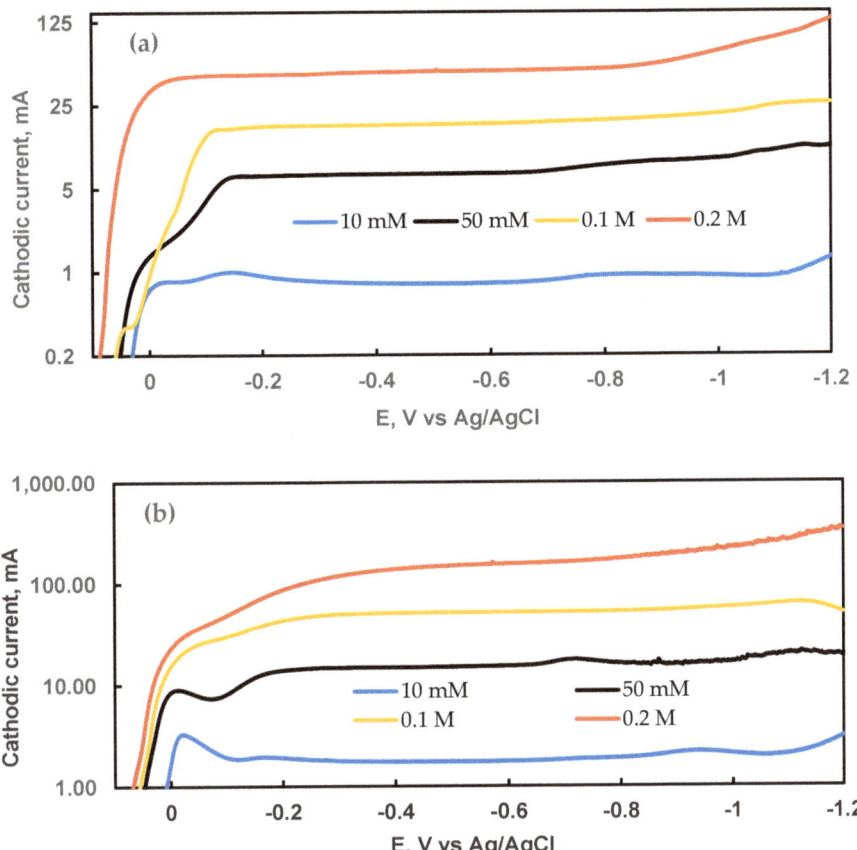

Figure 2. Cathodic voltammograms on Cu plate (**a**) and foam (**b**) obtained in the electrolytes with various concentrations of CuSO$_4$ (the compositions of solutions are shown in Table 1), potential scan rate 2 mV/s.

Table 2. Cu deposition rates on 2D and 3D electrodes in the solution containing 0.1 M CuSO$_4$ and 0.4 M Na$_2$SO$_4$.

Cu Plate			Cu Foam		
E, V versus Ag/AgCl	Deposition Time (s)	Cu Deposition Rate (mg/min)	E, V versus Ag/AgCl	Deposition Time (s)	Cu Deposition Rate (mg/min)
−0.1	1763	0.33	−0.1	643	0.94
−0.2	1681	0.35	−0.2	574	1.1
−0.4	1603	0.36	−0.4	593	1.0
−0.6	1571	0.37	−0.6	518	1.2

Chronoamperometric measurements (Table 2) clearly show an approximately 3 times faster copper deposition rate on the foam at all tested potentials. In this case, there was no hydrogen evolution, and the deposition efficiency was almost 100% on both substrates. A considerably higher deposition rate on the cooper foam substrate supports the idea that the deposition is controlled by diffusion to the electrode having a higher specific surface area. In addition, a higher metal deposition rate on the foam electrodes makes them an attractive substrate for the electrowinning of metals compared to other materials having a similar geometric area. The morphology of deposits is influenced by the potential and type of substrate, as it is shown in the SEM images in Figure 3.

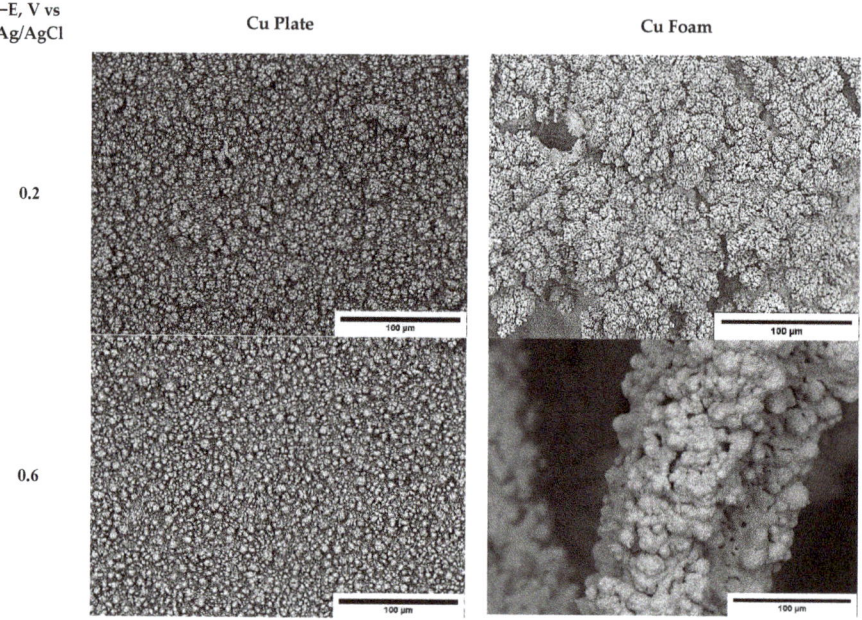

Figure 3. SEM images of potentiostatically electrodeposited Cu coatings at different cathodic potentials on flat and foam copper substrates after 30 C passed charge. The bath was 0.1 M $CuSO_4$ and 0.4 M Na_2SO_4.

The copper deposits have globules shapes on the flat electrodes, and the morphology did not differ at these two potentials. This is related to the very similar electrochemical deposition rates at these potentials, and as it can be seen from the voltammetry data (Figure 2) and efficiency of deposition, there were no side reactions, and the current was similar at these two potentials. Another case is the deposition on the porous substrate. At −0.2 V versus Ag/AgCl, copper forms cauliflower-like crystalline agglomerates with well-defined edges. At higher potential, the copper forms smoother surfaces that are still cauliflower-like structures. The coverage of both surface geometries was good even without external agitation, even at low potentials.

3.2. Surface Area and Diffusion Rate Estimations

To characterize copper foams and estimate the active surface areas for the charge and mass transfer processes that occur during the electrochemical deposition of copper, we utilized the EIS technique. EIS measurements have been done for all the solutions listed in Table 1. EIS measurements were performed at cathodic potentials of −0.125, −0.15, −0.175, and −0.2 V versus Ag/AgCl on flat and porous copper substrates. These potentials were chosen based on chronoamperometric data. At such low potentials, the change of surface morphology during deposition is still minimal and can be ignored in this case. Typical EIS scans on the copper plate at various potentials are shown in Figure 4. From the EIS data plots, we can see that at investigated potentials, the data plot can be divided into two zones: the high-frequency semicircle and the low-frequency (starting around 75–100 Hz) 45° angle line. The high-frequency semicircle can be attributed to charge up of the double layer and charge transfer to the copper ions, whilst the low-frequency line is attributed to the formation of the concentration gradient of the copper ions. To better evaluate ongoing processes, EIS data were fitted to the equivalent electric circuit (EEC) that is shown as an inset in Figure 4 of the Nyquist plot (a). The elements of applied EEC have the following physical meaning: R0 is resistance at the electrode/electrolyte interface, CPE(DL) is a double-layer capacitance modeled via the constant phase element (CPE), R(CT) is a charge transfer resistance, CPE(W) stands for the capacitance caused by the concentration gradient, and R(Diff) is a resistance caused by the concentration gradient. The element CPE(W) is attributed to the diffusion because of the signature 45° angle seen in the Nyquist plots at low frequencies (Figure 4), and the value n in this CPE element was very close to 0.5 in all the experiments. This constant phase element acting only in the low-frequency region represents diffusion, and it can be used as a Warburg element when $n = 0.5$ [48,49]. The values of the constant phase element CPE(DL) have been recalculated into true capacitance using Hsu and Mansfeld's equation [50]. All values of components of the fitted EEC are indicated in Table 3.

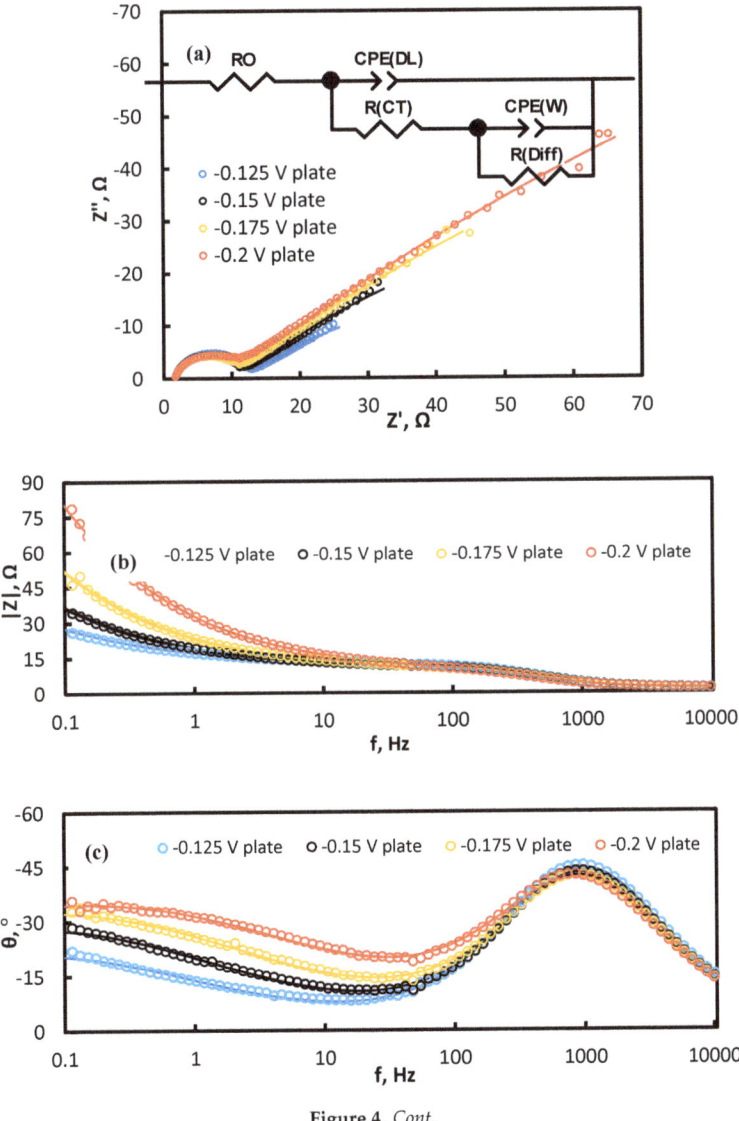

Figure 4. Cont.

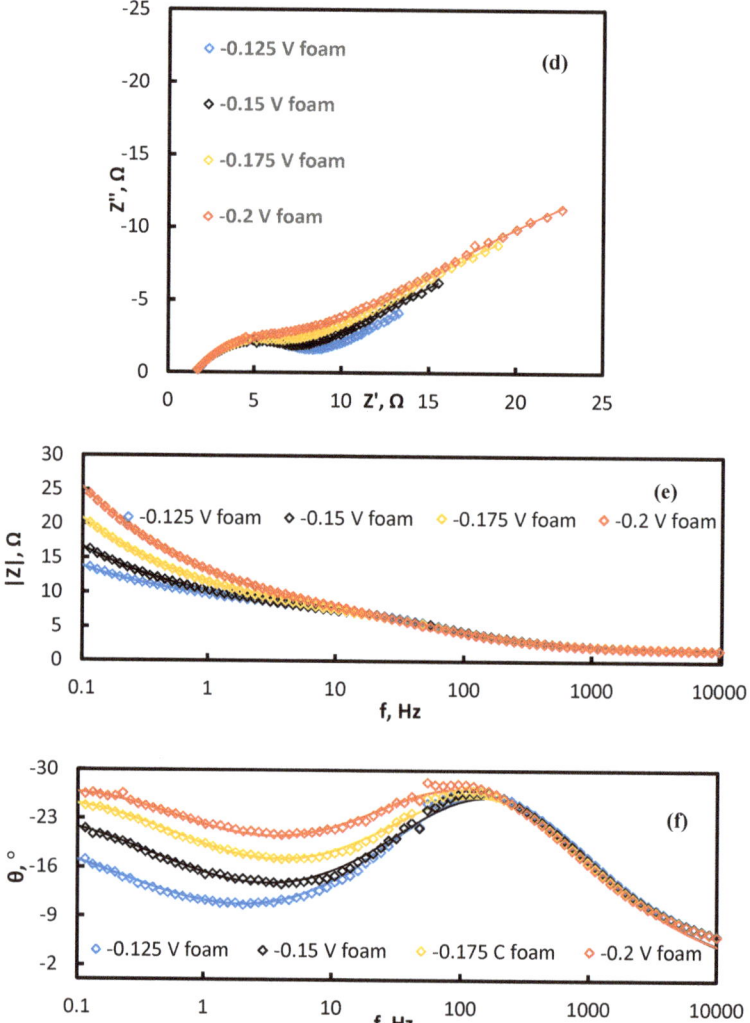

Figure 4. Nyquist (**a,d**) and Bode plots (**b–c,e–f**) on Cu plate (**a–c**) and foam (**d–f**) registered at various potentials (indicated on graphs) in 0.1 M CuSO$_4$ + 0.4 M Na$_2$SO$_4$ solution at 20 °C. Points—experimental data, solid lines—results of fitting to equivalent electric circuit (EEC) shown in the inset (**a**).

As it is seen, the proposed EEC describes well experimental EIS data on both substrates in a whole investigated potential range. The values of the capacitance of the double electric layer on both substrates might be used to estimate differences in real areas between the plate and foam electrodes, i.e., to estimate the roughness factor as a ratio of C(DL) on foam and plates that have the same geometric area (1 cm × 1 cm). Notably, the double-layer capacitance (C(DL)) extracted from the EIS data is 50 µF (see Figure 5), and it is in good agreement with the theoretical values assigned to 1 cm^2 of copper [49]. The capacitance of the double layer of a commercial foam, that has the same geometric area as a plate, is 7 to 14 times higher in comparison with a plate electrode. The thickness of the double electric layer is very small and is in tens of nanometers; therefore, this layer replicates the surface morphology on the nano-level, and the ratio with the value obtained on the plate electrode can represent the roughness factor, and it matches the ratio of C(DL) of both surfaces –(C(DLfoam); C(DL plate) is 7–14:1). However, the increase of double-layer capacitances with the increase of applied cathodic potential on both flat and porous surfaces is different. On the porous electrode, the C(DL) increase is much higher when compared to the change in capacitances of the flat electrode. This increase is related to the much higher surface area, and the distribution of current on the surface of the foam. With higher potential, the current distributes more evenly on the whole foam surface, and the edge effect is less apparent, which also influences the surface area estimations [51,52].

When looking at the effect that the concentration of copper ions has on the EIS parameters (Table 3), we can divide the results into three sections: high concentration (0.2 M), mid-level concentrations (0.1 and 0.05 M), and low concentrations (0.01 M). The double electric layer (DL) capacitance values do not differ that much with the change of the concentration on both surface geometries. However, when looking at charge transfer resistance, the differences between concentrations are significant. At low concentrations, the charge transfer resistance is very high; this is caused by the lack of copper ions. In contrast, this resistance at mid-level concentrations is around 6–9 Ω, which depends on the surface geometry as well as applied potential (Figure 5). At high concentrations (0.2 M and higher), the charge transfer resistance values decrease approximately 3 times on both surfaces, because of an abundance of conducting particles. Nevertheless, this charge transfer resistance is lower at all investigated potentials and all concentrations on the foam electrode, showing that the reduction reaction occurs faster on the copper foams.

When taking a look at the charge transfer resistance dependence on potential (Figure 5) with both types of electrodes, it is clear that the 3D electrode displays approximately 1.5–1.7 times lower charge transfer resistance than the 2D electrode, agreeing with the results of voltammetry (see Figure 2). The differences in the charge transfer resistance on plate and foam electrodes are lower than the differences in the capacitances of DL, because the reaction layer is thicker than the DL, and in some areas of the foam electrode, it overlaps. As it can be seen from Figure 5, the difference between 2D and 3D electrodes in charge transfer resistance is higher at low potentials; thus, the charge transfer reaction on the foam occurs easier, and it partially explains the higher Cu deposition rate (see Table 2). However, lowering the charge transfer resistance, or in turn, the increase of the rate of the charge transfer reaction by approximately 2 times, does not result in increases in the Cu deposition rate by approximately 3 times.

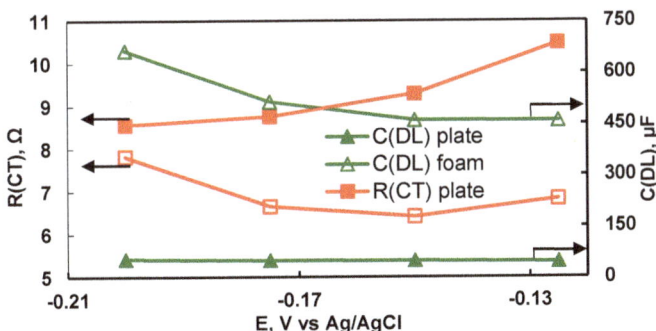

Figure 5. Dependence of double-layer capacitance (ordinate at the right) and charge transfer resistance (ordinate at the left) on potential applied for Cu plate and foam electrodes in 0.1 M $CuSO_4$ + 0.4 M Na_2SO_4 solution.

To further characterize the difference in copper deposition reactions on flat and porous copper surfaces, the components of EEC related to diffusion have been investigated in detail (Figure 6). The foam has lower charge transfer resistance, meaning faster reactions and better hydrodynamic qualities, allowing for faster diffusion and in turn the much faster deposition, even with a larger surface and in turn, lower current density.

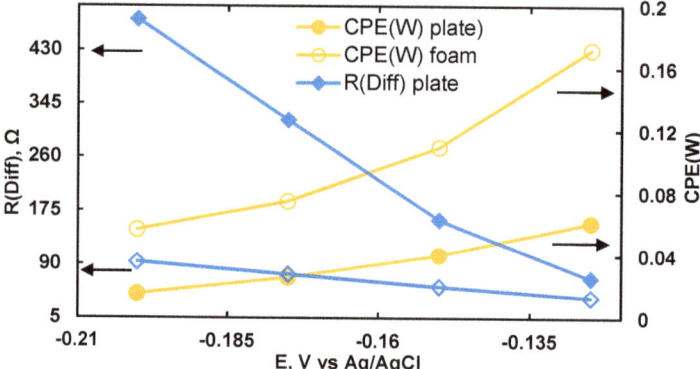

Figure 6. Dependence of diffusion-related elements of EEC on the potential applied. Measurements performed using a copper plate and copper foam as working electrodes in 0.1 M $CuSO_4$ + 0.4 M Na_2SO_4 solution.

Table 3. Values of electrochemical impedance spectroscopy (EIS) parameters obtained by fitting data obtained on copper foam and copper plates at −0.175 V versus Ag/AgCl at different copper concentrations. EC used for modeling shown in Figure 3 inset. CPE(DL): a double-layer (DL) capacitance modeled via the constant phase element (CPE), CPE(W): the capacitance caused by the concentration gradient, R(CT): charge transfer resistance, R(Diff): resistance caused by the concentration gradient.

Cu Plate	0.2 M $CuSO_4$ + 0.3 M Na_2SO_4	0.1 M $CuSO_4$ + 0.4 M Na_2SO_4	0.05 M $CuSO_4$ + 0.45 M Na_2SO_4	0.01 M $CuSO_4$ + 0.49 M Na_2SO_4
C(DL), µF	40.5	49.3	41.9	56.4
R(CT), Ω	2.98	8.76	7.44	54.66
CPE(W)	0.0696	0.0260	0.0285	0.00247
R(Diff), Ω	233.4	319.3	199.6	663.7
Cu Foam			−	
C(DL), µF	299.2	513.5	456.6	754.1
R(CT), Ω	2.39	6.65	6.74	122.20
CPE(W)	0.2033	0.0748	0.0668	0.0039
R(Diff), Ω	14.9	74.9	83.3	1551.0

The parameter related to diffusion CPE(W) at low concentrations is almost equal on both surface geometries, showing that the diffusion effect is similar, but the resistance at low concentration is about 2.5 times higher. It means that the diffusion layer is much thicker on the copper foams surface because of the porosity effect. Therefore, it causes a higher rate of copper electrodeposition. The overall trend in mid-level and high concentrations is that with the increase of Cu^{2+} concentration, the CPE(W) value increases, and the R(Diff) decreases. As it is seen from Table 3, the difference between R(Diff) values at 0.2 and 0.05 M concentrations on the flat surface is only around 14%, whereas on the foam electrode, the values of R(Diff) are lower, but all values are sensitive to the concentration of Cu(II) in the solution. The highest value of R(Diff) is obtained on the foam electrode at a relatively low concentration of Cu(II), i.e., 0.01 M, which is probably due to the faster depletion of copper ion concentration in the 3D diffusion layer and the necessity of a longer time to supply Cu(II) ions into the pores. Since the deposition rate on the foam electrode at a higher concentration of Cu(II) is 3 times faster than on the flat electrode, this is mirrored by the behavior of CPE(W), showing that the diffusion occurs 3 times faster on the foam. The efficiency of charge transfer on the porous surfaces is higher as well, which is in good agreement with other studies of metal depositions on porous surfaces [31].

To even better understand the diffusion peculiarities on 2D and 3D electrodes, the diffusion impedance using extracted values from total impedance data (presented in Table 4) was calculated. As it is shown in Figure 4, the copper deposition occurs under diffusion control at low frequencies (below 100 Hz) on both foam and plate electrodes, and diffusion is modeled by a parallel connection of CPE(W) and R(Diff) elements (see Figure 4). In this case, diffusion impedance, Z_{diff}, as a function of frequency is calculated by the equation:

$$Z_{diff}(\omega) = \frac{R_{Diff}}{1 + (j\omega)^{\alpha} Q R_{Diff}} \quad (1)$$

where Q and α are parameters of CPE(W), R(Diff) is resistance caused by diffusion, and ω is the phase angle ($\omega = 2\pi f$). However, when $\alpha = 1 - Q$ is pure capacitance, in our case, $\alpha = 0.5$, and the CPE represents diffusion [53].

The calculated diffusion impedance data are presented in Figure 7. As it is seen, the diffusion impedance on the plate Cu electrode is 2–4 times higher than that on the foam Cu electrodes, which is dependent on the frequency and potential applied.

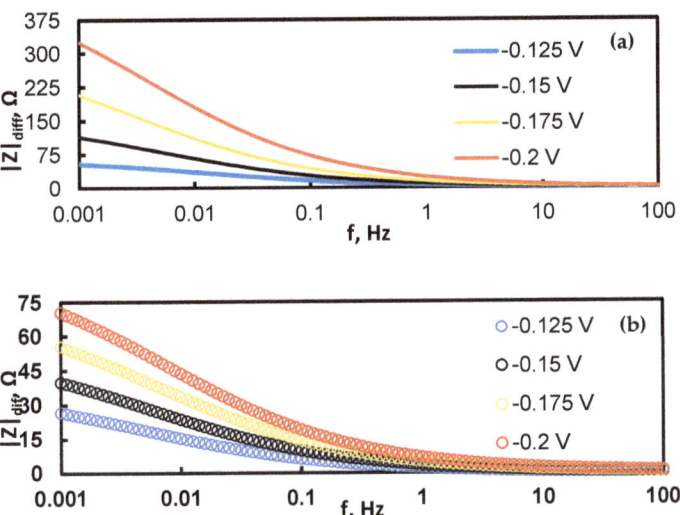

Figure 7. Bode plots of extracted diffusion impedance at various potentials on flat Cu substrate (**a**); and Cu foam substrate (**b**).

These results once again confirm the chronopotentiometric data obtained on both 2D and 3D Cu electrodes. For chronopotentiometry experiments, current values have been chosen higher than the limiting current values seen in Figure 8. In this case, the transition time at which the concentration of metal ions on the electrode becomes equal to zero is visual on the chronopotentiograms, and the effective diffusion coefficient can be calculated by the Sand equation:

$$i\sqrt{\tau} = \frac{nFAC_0 \sqrt{\pi D_{eff}}}{2} \quad (2)$$

where τ is a transition time (s), i is a current (A), C_0 is the concentration of Cu(II) ions (mol/cm^3), D_{eff} is the effective diffusion coefficient (cm$^2 \cdot$s^{-1}), F is Faraday's constant, n is the number of electrons participating in the electrochemical reaction; and A is a geometrical surface area.

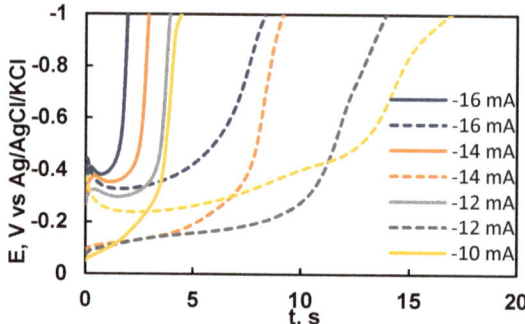

Figure 8. Chronopotentiograms on flat (continuous lines) and porous (dashed lines) electrodes at various current densities in 50 mM CuSO$_4$ and 0.45 M Na$_2$SO$_4$ solution. All the densities have been calculated for the geometrical area of the substrate of 1 cm^2.

In our case $i\sqrt{\tau} \sim const$, so the maximal deposition rate is controlled by the mass transfer. The values of the effective diffusion coefficient of Cu^{2+} ions on both plate and foam Cu electrodes were calculated by Equation (2), and the data are shown in Table 4. The effective diffusion coefficient on the plate electrode is almost three times lower than on the foam electrode, and it is in good agreement with EIS data.

Table 4. Effects of electrode geometry on effective Cu(II) ions diffusion coefficient.

Applied Current	Effective Diffusion Coefficient	
	Plate	Foam
I, mA	10^6 D, cm$^2 \cdot$s^{-1}	10^6 D, cm$^2 \cdot$s^{-1}
−10	6.79	18.06
−12	6.72	19.70
−14	6.73	20.16
−16	6.62	20.77
Average D_{eff}	6.72	19.67

So, copper foams are great substrates for reactions that are either limited by the mass transfer (electrochemical depositions, etc.) or the ones that are restricted by adsorption or activation (HER and similar), making them great candidates to reduce the size of electrodes, but not to lose out on the efficiency and activity of electrodes.

4. Conclusions

A comprehensive investigation of the electrochemical deposition of copper onto 2D (plate) and 3D (foam) Cu substrates has been done. Using various electrochemical methods, it was determined that the rate-determining step in a copper deposition is diffusion. The main processes occurring on the electrode are the charge-up of double electric layer, charge transfer, and diffusion. The specific electrochemically active area of Cu foam was estimated from EIS data, and based on the values of the double electric layer, it was determined to be 7–14 times higher than that for the plate electrode. Based on the EIS data, it was determined that the charge transfer resistance on the Cu foam electrode is 1.5–1.7 times lower than that on the Cu plate electrode, which results in an increase in a charge transfer rate of approximately 2 times. Based on the analysis of the diffusion impedance and chronopotentiometry data, it was found that Cu^{2+} mass transfer and the copper deposition rate is up to 3 times faster on the foam surface in comparison with a flat surface having the same geometric area in the same potential range. In addition, effective diffusion coefficients have been calculated from chronopotentiometry data using Sand's equation. These findings make Cu foam an attractive material for metal electrowinning processes as well as for processes controlled by adsorption (e.g., hydrogen evolution reaction).

Author Contributions: Investigation, M.V. and N.T.; methodology, M.V. and H.C.; supervision, H.C. and N.T.; visualization, M.V.; writing—original draft preparation, M.V.; writing—review and editing, M.V., H.C. and N.T. All authors have read and agreed to the published version of the manuscript.

Funding: This work was funded by the Lithuanian Business Support Agency (LVPA); project J05-LVPA-K-01-0022.

Conflicts of Interest: The authors declare no conflict of interest.

References

1. Liu, X.; Tanaka, M.; Matsui, Y. Generation amount prediction and material flow analysis of electronic waste: A case study in Beijing, China. *Waste Manag. Res.* **2006**, *24*, 434–445. [CrossRef] [PubMed]
2. Jain, A.; Sareen, R. E-waste assessment methodology and validation in India. *J. Mater. Cycles Waste Manag.* **2006**, *8*, 40–45. [CrossRef]
3. Eurostat, Statistics Explained. Available online: https://ec.europa.eu/eurostat/statistics-explained/index.php/Waste_statistics_-_electrical_and_electronic_equipment (accessed on 7 August 2020).
4. Vegliò, F.; Quaresima, R.; Fornari, P.; Ubaldini, S. Recovery of valuable metals from electronic and galvanic industrial wastes by leaching and electrowinning. *Waste Manag.* **2003**, *23*, 245–252. [CrossRef]
5. Grimshaw, P.; Calo, J.M.; Hradil, G. Cyclic electrowinning/precipitation (CEP) system for the removal of heavy metal mixtures from aqueous solutions. *Chem. Eng. J.* **2011**, *175*, 103–109. [CrossRef] [PubMed]
6. Bertuol, D.A.; Amado, F.D.R.; Veit, H.; Ferreira, J.Z.; Bernardes, A.M. Recovery of nickel and cobalt from spent nimh batteries by electrowinning. *Chem. Eng. Technol.* **2012**, *35*, 2084–2092. [CrossRef]
7. Kim, J.H.; Kim, R.H.; Kwon, H.S. Preparation of copper foam with 3-dimensionally interconnected spherical pore network by electrodeposition. *Electrochem. Commun.* **2008**, *10*, 1148–1151. [CrossRef]
8. Niu, J.; Liu, X.; Xia, K.; Xu, L.; Xu, Y.; Fang, X.; Lu, W. Effect of electrodeposition parameters on the morphology of three-dimensional porous copper foams. *Int. J. Electrochem. Sci.* **2015**, *10*, 7331–7340.
9. Shahbazi, P.; Kiani, A. Fabricated Cu_2O porous foam using electrodeposition and thermal oxidation as a photocatalyst under visible light toward hydrogen evolution from water. *Int. J. Hydrogen Energy* **2016**, *41*, 17247–17256. [CrossRef]
10. Shin, H.C.; Dong, J.; Liu, M. Nanoporous structures prepared by an electrochemical deposition process. *Adv. Mater.* **2003**, *15*, 1610–1614. [CrossRef]
11. Zhang, W.; Ding, C.; Wang, A.; Zeng, Y. 3-D Network pore structures in copper foams by electrodeposition and hydrogen bubble templating mechanism. *J. Electrochem. Soc.* **2015**, *162*, D365–D370. [CrossRef]
12. Nam, D.; Kim, R.; Han, D.; Kim, J.; Kwon, H. Effects of $(NH_4)_2SO_4$ and BTA on the nanostructure of copper foam prepared by electrodeposition. *Electrochim. Acta* **2011**, *56*, 9397–9405. [CrossRef]
13. Li, D.; Podlaha, E.J. Template-assisted electrodeposition of porous Fe–Ni–Co nanowires with vigorous hydrogen evolution. *Nano Lett.* **2019**, *19*, 3569–3574. [CrossRef] [PubMed]
14. Raoof, J.B.; Ojani, R.; Kiani, A.; Rashid-Nadimi, S. Fabrication of highly porous Pt coated nanostructured Cu-foam modified copper electrode and its enhanced catalytic ability for hydrogen evolution reaction. *Int. J. Hydrogen Energy* **2010**, *35*, 452–458. [CrossRef]
15. Luo, Z.-H.; Feng, M.; Lu, H.; Kong, X.-X.; Cao, G.-P. Nitrile butadiene rubber hydrogenation over a monolithic Pd/CNTs@Nickel foam catalysts: Tunable CNTs morphology effect on catalytic performance. *Ind. Eng. Chem. Res.* **2019**, *58*, 1812–1822. [CrossRef]
16. Ashby, M.F.; Evans, A.; Fleck, N.A.; Gibson, L.J.; Hutchinson, J.W.; Wadley, H.N. Metal foams: A design guide. *Mater. Des.* **2002**, *23*, 119. [CrossRef]
17. Abdel-karim, R.; El-raghy, S. Electrochemical Deposition of Nanoporous Metallic Foams for Energy Applications. pp. 69–91. Available online: http://www.onecentralpress.com/wp-content/uploads/2017/08/Chapter-4-AMA-.pdf (accessed on 22 August 2020).

18. Eugénio, S.; Demirci, U.B.; Silva, T.M.; Carmezim, M.J.; Montemor, M.F. Copper-cobalt foams as active and stable catalysts for hydrogen release by hydrolysis of sodium borohydride. *Int. J. Hydrogen Energy* **2016**, *41*, 8438–8448. [CrossRef]
19. Liu, W.; Hu, E.; Jiang, H.; Xiang, Y.; Weng, Z.; Li, M.; Fan, Q.; Yu, X.; Altman, E.I.; Wang, H. A highly active and stable hydrogen evolution catalyst based on pyrite-structured cobalt phosphosulfide. *Nat. Commun.* **2016**, *7*, 10771. [CrossRef]
20. Lange, G.A.; Eugénio, S.; Duarte, R.G.; Silva, T.M.; Carmezim, M.J.; Montemor, M.F. Characterisation and electrochemical behaviour of electrodeposited Cu–Fe foams applied as pseudocapacitor electrodes. *J. Electroanal. Chem.* **2015**, *737*, 85–92. [CrossRef]
21. Murakami, T.; Akagi, T.; Kasai, E. Development of porous iron based material by slag foaming and its reduction. *Procedia Mater. Sci.* **2014**, *4*, 30–35. [CrossRef]
22. Kelpšaite, I.; Baltrušaitis, J.; Valatka, E. Electrochemical deposition of porous cobalt oxide films on AISI 304 type steel. *Medziagotyra* **2011**, *17*, 236–243.
23. Mattarozzi, L.; Cattarin, S.; Comisso, N.; Gerbasi, R.; Guerriero, P.; Musiani, M.; Vazquez-Gomez, L.; Verlato, E. Electrodeposition of Cu-Ni alloy electrodes with bimodal porosity and their use for nitrate reduction. *ECS Electrochem. Lett.* **2013**, *2*, D58–D60. [CrossRef]
24. Rehman, T.U.; Ali, H.M.; Saieed, A.; Pao, W.; Ali, M. Copper foam/PCMs based heat sinks: An experimental study for electronic cooling systems. *Int. J. Heat Mass Transf.* **2018**, *127*, 381–393. [CrossRef]
25. Zhao, J.; Zou, X.; Sun, P.; Cui, G. Three-dimensional Bi-continuous nanoporous Gold/Nickel foam supported MnO_2 for high performance supercapacitors. *Sci. Rep.* **2017**, *7*, 1–8. [CrossRef] [PubMed]
26. Liu, Y.; Hangarter, C.M.; Garcia, D.; Moffat, T.P. Self-terminating electrodeposition of ultrathin Pt films on Ni: An active, low-cost electrode for H2 production. *Surf. Sci.* **2015**, *631*, 141–154. [CrossRef]
27. Vainoris, M.; Tsyntsaru, N.; Cesiulis, H. Modified electrodeposited cobalt foam coatings as sensors for detection of free chlorine in water. *Coatings* **2019**, *9*, 306. [CrossRef]
28. Ma, S.; Zhou, H.C. Gas storage in porous metal-organic frameworks for clean energy applications. *Chem. Commun.* **2010**, *46*, 44–53. [CrossRef] [PubMed]
29. Karimi Shervedani, R.; Lasia, A. Evaluation of the surface roughness of microporous Ni–Zn–P electrodes by in situ methods. *J. Appl. Electrochem.* **1999**, *29*, 979–986. [CrossRef]
30. Popov, K.I.; Nikolić, N.D.; Živković, P.M.; Branković, G. The effect of the electrode surface roughness at low level of coarseness on the polarization characteristics of electrochemical processes. *Electrochim. Acta* **2010**, *55*, 1919–1925. [CrossRef]
31. Miranda-Hernández, M.; González, I.; Batina, N. Silver electrocrystallization onto carbon electrodes with different surface morphology: Active sites vs surface features. *J. Phys. Chem. B* **2001**, *105*, 4214–4223. [CrossRef]
32. Menshykau, D.; Streeter, I.; Compton, R.G. Influence of electrode roughness on cyclic voltammetry. *J. Phys. Chem. C* **2008**, *112*, 14428–14438. [CrossRef]
33. Kostevšek, N.; Rožman, K.Ž.; Pečko, D.; Pihlar, B.; Kobe, S. 33NN A comparative study of the electrochemical deposition kinetics of iron-palladium alloys on a flat electrode and in a porous alumina template. *Electrochim. Acta* **2014**, *125*, 320–329. [CrossRef]
34. Gira, M.J.; Tkacz, K.P.; Hampton, J.R. Physical and electrochemical area determination of electrodeposited Ni, Co, and NiCo thin films. *Nano Converg.* **2015**, *3*, 6. [CrossRef] [PubMed]
35. Zankowski, S.P.; Vereecken, P.M. Electrochemical determination of porosity and surface area of thin films of interconnected nickel nanowires. *J. Electrochem. Soc.* **2019**, *166*, D227–D235. [CrossRef]
36. Tadros, T.F.; Lyklema, J. Adsorption of potential—determining ions at the silica—aqueous electrolyte interface and the role of some cations. *J. Electroanal. Chem. Interf. Electrochem.* **1968**, *17*, 267–275. [CrossRef]
37. Schneider, I.A.; Kramer, D.; Wokaun, A.; Scherer, G.G. Effect of inert gas flow on hydrogen underpotential deposition measurements in polymer electrolyte fuel cells. *Electrochem. Commun.* **2007**, *9*, 1607–1612. [CrossRef]
38. Green, C.L.; Kucernak, A. Determination of the platinum and ruthenium surface areas in platinum-ruthenium alloy electrocatalysts by underpotential deposition of Copper. I. Unsupported catalysts. *J. Phys. Chem. B* **2002**, *106*, 1036–1047. [CrossRef]

39. Yamaguchi, R.; Kurosu, S.; Suzuki, M. Hydroxyl radical generation by zero-valent iron/Cu (ZVI/Cu) bimetallic catalyst in wastewater treatment: Heterogeneous Fenton/Fenton-like reactions by Fenton reagents formed in-situ under oxic conditions. *Chem. Eng. J.* **2018**, *334*, 1537–1549. [CrossRef]
40. Macht, F.; Eusterhues, K.; Pronk, G.J.; Totsche, K.U. Specific surface area of clay minerals: Comparison between atomic force microscopy measurements and bulk-gas (N_2) and -liquid (EGME) adsorption methods. *Appl. Clay Sci.* **2011**, *53*, 20–26. [CrossRef]
41. Sharifi-Viand, A.; Mahjani, M.G.; Jafarian, M. Determination of fractal rough surface of polypyrrole film: AFM and electrochemical analysis. *Synth. Met.* **2014**, *191*, 104–112. [CrossRef]
42. Wongmanerod, C.; Zangooie, S.; Arwin, H. Determination of pore size distribution and surface area of thin porous silicon layers by spectroscopic ellipsometry. *Appl. Surf. Sci.* **2001**, *172*, 117–125. [CrossRef]
43. Damian, A.; Omanovic, S. Ni and Ni{single bond}Mo hydrogen evolution electrocatalysts electrodeposited in a polyaniline matrix. *J. Power Sources* **2006**, *158*, 464–476. [CrossRef]
44. Kandalkar, S.G.; Lee, H.M.; Chae, H.; Kim, C.K. Structural, morphological, and electrical characteristics of the electrodeposited cobalt oxide electrode for supercapacitor applications. *Mater. Res. Bull.* **2011**, *46*, 48–51. [CrossRef]
45. Song, H.; Song, H.; Jung, Y.; Jung, Y.; Lee, K.; Lee, K.; Dao, L.H.; Dao, L.H. Electrochemical impedance spectroscopy of porous electrodes: The effect of pore size distribution. *Electrochim. Acta* **1999**, *44*, 3513–3519. [CrossRef]
46. Ogihara, N.; Itou, Y.; Sasaki, T.; Takeuchi, Y. Impedance spectroscopy characterization of porous electrodes under different electrode thickness using a symmetric cell for high-performance lithium-ion batteries. *J. Phys. Chem. C* **2015**, *119*, 4612–4619. [CrossRef]
47. Yan, B.; Li, M.; Li, X.; Bai, Z.; Dong, L.; Li, D. Electrochemical impedance spectroscopy illuminating performance evolution of porous core-shell structured nickel/nickel oxide anode materials. *Electrochim. Acta* **2015**, *164*, 55–61. [CrossRef]
48. Kaufmann, B.Y.K. Transfer function simulation for electrochemical impedance spectroscopy (EIS). *Rev. Colomb. Fis.* **2005**, *37*, 25–27.
49. Mahato, N.; Singh, M.M. Investigation of passive film properties and pitting resistance of AISI 316 in aqueous ethanoic acid containing chloride ions using electrochemical impedance spectroscopy(EIS). *Port. Electrochim. Acta* **2011**, *29*, 233–251. [CrossRef]
50. Hsu, C.H.; Mansfeld, F. Concerning the conversion of the constant phase element parameter Y_0 into a capacitance. *Corrosion* **2001**, *57*, 747–748. [CrossRef]
51. Krzewska, S. Impedance investigation of the mechanism of copper electrodeposition from acidic perchlorate electrolyte. *Electrochim. Acta* **1997**, *42*, 3531–3540. [CrossRef]
52. Halsey, T.C. Frequency dependence of the double-layer impedance at a rough surface. *Phys. Rev. A* **1987**, *35*, 3512–3521. [CrossRef]
53. Hirschorn, B.; Orazem, M.E.; Tribollet, B.; Vivier, V.; Frateur, I.; Musiani, M. Determination of effective capacitance and film thickness from constant-phase-element parameters. *Electrochim. Acta* **2010**, *55*, 6218–6227. [CrossRef]

© 2020 by the authors. Licensee MDPI, Basel, Switzerland. This article is an open access article distributed under the terms and conditions of the Creative Commons Attribution (CC BY) license (http://creativecommons.org/licenses/by/4.0/).

Article

The Mechanisms of Degradation of Titanium Dental Implants

Agnieszka Ossowska * and Andrzej Zieliński

Department of Materials Engineering and Bonding, Gdańsk University of Technology, 80233 Gdańsk, Poland; azielins@pg.edu.pl
* Correspondence: agnieszka.ossowska@pg.edu.pl; Tel.: +48-58-347-19-63

Received: 29 July 2020; Accepted: 26 August 2020; Published: 28 August 2020

Abstract: Titanium dental implants show very good properties, unfortunately there are still issues regarding material wear due to corrosion, implant loosening, as well as biological factors—allergic reactions and inflammation leading to rejection of the implanted material. In order to avoid performing reimplantation operations, changes in the chemical composition and/or modifications of the surface layer of the materials are used. This research is aimed at explaining the possible mechanisms of titanium dissolution and the role of oxide coating, and its damage, in the enhancement of the corrosion process. The studies of new and used implants were made by scanning electron microscopy and computer tomography. The long-term chemical dissolution of rutile was studied in Ringer's solution and artificial saliva at various pH levels and room temperature. Inductively coupled plasma mass spectrometry (ICP-MS) conjugated plasma ion spectrometry was used to determine the number of dissolved titanium ions in the solutions. The obtained results demonstrated the extremely low dissolution rate of rutile, slightly increasing along with pH. The diffusion calculations showed that the diffusion of titanium through the oxide layer at human body temperature is negligible. The obtained results indicate that the surface damage followed by titanium dissolution is initiated at the defects caused by either the manufacturing process or implantation surgery. At a low thickness of titanium oxide coating, there is a stepwise appearance and development of cracks that forms corrosion tunnels within the oxide coating.

Keywords: dental implants; corrosion; ringer's solution; artificial saliva; titanium oxide layers; inductively coupled plasma mass spectrometry (ICP-MS)

1. Introduction

Titanium alloys possess good strength properties and high resistance to the most aggressive environments such as hydrochloric acid or sulfuric acid [1–3]. The compact, stable oxide layer [4] is responsible for corrosion resistance and biotolerance, effectively stopping the anodic pickling of the substrate [5–7]. Another important function is to chemically stabilize the implant in the living organism [4,8–10]. The more compact and bonded the passive layer is to the substrate, the better the corrosion resistance. In the case of thick oxide layers, an improvement in tribological properties may also be observed [7]. According to Hanawa et al. [11], the top sublayer of the titanium oxide layer inhibits metal ion release [12,13] and its transformation in vitro. Additionally, the oxide layer promotes osseointegration and bone adhesion [14–16].

It is known that each implant inserted into the body is treated as a foreign body and can cause allergic reactions, inflammations, even the rejection of the implant. The human body is a very specific environment as the body fluids—extracellular fluids and blood—contain aqueous solutions of certain organic substances, dissolved oxygen, various inorganic anions (Cl^-, HPO_4^{2-}, HCO^{3-}), and cations (Na^+, K^+, Ca^{2+}, Mg^{2+}), which together represent a highly aggressive environment [17]. The presence of amino acids and proteins accelerates the corrosion processes [18]. Besides, in the case of dental implants, the composition of the saliva is highly complex, containing both inorganic salts and organic components. This composition depends on many factors such as food, age, and diseases and the pH of saliva can vary around dental implants. The ingestion of acidic beverages can decrease the buccal pH, and the infections can also acidify the pH of saliva, contributing to the corrosion of dental implants. On the other hand, titanium and its alloys are sensitive to tribocorrosion [19]. The oxidation is performed mainly to prevent corrosion of titanium and its alloys in severe conditions. For dental implants, such oral environments include varying pH, acid attack and the presence of chemical compounds such as cetylpyridinium chloride, sodium fluoride and hydrogen peroxide [20].

The action of media containing fluoride ions causes degradation of the continuity of the oxide film followed by damage to the titanium as a result of the ingestion of fluoride ions into the oxide layer, thereby reducing its protective properties [21,22].

Corrosion processes influence changes in the structure of the implanted material, weakening its integrity, which can result in material discontinuities and cracks. Cells in direct contact with the exposed surface of the material are stimulated for the intensive secretion of inflammatory mediators, mainly neutrophils and macrophages [23]. In vitro studies [24] show that corrosion products are harmful to cell differentiation and proliferation processes.

The dissolution of the titanium oxide layer is due to the process of ion diffusion into the layer. In titanium, the oxygen atoms migrate via the interstitial diffusion mechanism, occupying the free, octahedral interstitial positions in the titanium hexagonal lattice. Studies conducted by Wu and Trinkle [25] showed that for oxygen atoms not only interstitial but also axial positions are available, i.e., all arrangements of oxygen atoms in the titanium matrix are possible.

The oxidation of titanium is faster when the material is subjected to high temperatures and the influence of an oxygenated environment. The overall oxidation reaction includes the formation of oxide followed by the diffusion of oxygen into the bulk of the titanium. Oxygen diffusion creates an oxygen-enriched layer due to the high solubility of oxygen in the titanium and the oxygen stabilizing effect in the crystalline titanium structure [26]. In solids, the most likely atomic diffusion mechanism is a vacancy or interstitial mechanism, i.e., the motion of atoms occurs as the consequence of the presence of imperfections [27]. The interstitial diffusion mechanism is typical for low atomic radius atoms such as hydrogen, oxygen, carbon, and nitrogen.

The great advantage of the oxide layers produced on titanium and its alloys is their capability of repassivation, and some released ions depend on regeneration. Hanawa et al. [11], while measuring repassivation potentials, estimated the recovery rate of the oxide layer in 0.9% physiological saline: for 316L steel as 35.3 min, for Ti6Al4V as 8.2 min, and Co28Cr6Mo as 12.7 min. The research conducted by Hanawa et al. [28] showed that in Hanks' solution, the rate of repassivation was lower than in 0.9% saline solution.

Metallic elements have a different tendency to release ions, and even trace amounts of elements in the alloy composition should not be neglected [11]. There are data on the significant contents of some alloying elements of Ti6Al4V within the tissue around the implanted alloy. So far [29] reports on the consequences of ion release into the body have focused on the importance of the possible impact of released ions on biomolecules and the initiation of adverse biological reactions as the titanium ions could quickly react with water molecules or inorganic anions, easily binding with body fluids.

Osseointegration involves a series of biological events influenced by multiple factors. Among them, the porous-structured Ti alloys have shown to allow rapid bone ingrowth and improved osseointegration by increasing the bone-implant interface area [30,31]. Such conditions are achieved for dental implants by micro-arc oxidation, which brings out the rough surface [19,32–35]. The bioactivity is usually increased by anodic oxidation in an electrolyte containing calcium phosphates [36,37], and wear resistance by incorporation of tough nanoparticles [35].

We have put a hypothesis that the damage of oxide coating can be sometimes or often, the main cause for the degradation of material and removal of the implant. The purpose of the study was to characterize the processes which allow for titanium dissolution from dental implants. To achieve that, surface examinations of new and applied dental implants were carried out. The dissolution rates of titanium dioxide (rutile) into two simulating body fluids at different pH values were performed. The titanium transport through the oxide coating was also calculated.

2. Materials and Methods

2.1. Microstructural Characterization of Surfaces of Implants

The first stage of research was the qualitative analysis of the surface as well as of the cross-sections for new and used (removed) dental implants. The tests were carried out on groups of samples:

- new dental implants in number of four, made of the Ti6Al4V alloy by four different companies (called as A, B, C, and D);
- used dental implants in number of fourteen, removed at the Warsaw Medical Academy, from the patients, no more than half a year after implantation, made of the Ti6Al4V alloy by four different companies (called as above).

The examinations of the surfaces and cross-sections of new and used dental implants were carried out at the Gdansk University of Technology using a scanning electron microscope (SEM; JEOL JSM-7600F, JEOL Ltd., Tokyo, Japan). Before observation the new and used surfaces of dental implants were cleaned in methanol. The cross-section samples were cut from the implants and were ground with abrasive papers (No. 2500 as the last, Struers Inc., Cleveland, OH, USA).

To obtain detailed information on the geometry and presence of cracks or delaminations of the coatings on the dental implants, a computer microtomography (CT) technique was used. The CT examinations were made with the µCT (General Electric, Lewistown, PA, USA) phoenix v-tome-x s using an X-ray "direct tube" with a set power of 17 W (70 kV, 100 µA). One thousand radiographs (2D X-rays) for each tomogram were made with 360° rotation and an exposure time of 333 ms (for a single radiogram). 3D tomograms were reconstructed from radiographs using the phoenix datos-x2 reconstruction program and a standard reconstruction algorithm. The reconstructed samples had a resolution of 2.413 µm/Voxel and were analyzed using the commercial VGStudio Max package.

2.2. Investigation of Dissolution Rate of Oxide Coatings

In the second stage, the tests of the dissolution rate of rutile (titanium oxide) were performed in two simulated body fluids (SBF). The starting material was the titanium oxide powder (purity of 98.0%–100%) delivered by Acros Organics (Morris Plains, NJ, USA). Cylindrical samples, of dimensions 6 mm × 3 mm (diameter × length), were prepared using the classical powder metallurgy method without a filler. The material was formed in a single-axis pressing process using a force of 2 kN acting on the stamp for 60 s at the position shown in Figure 1.

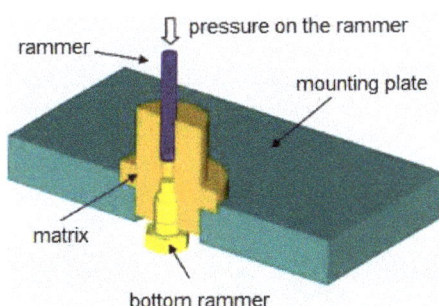

Figure 1. The cold pressing scheme for manufacturing the titanium dioxide specimens by powder metallurgy.

The sintering process was carried out in a chamber oven (Type 22 MRT/1300, Conbest Ltd., Kraków, Poland) at 1300 °C for 2 h in an air atmosphere. The samples were heated at a rate 0.5 °C/min up to 100 °C and then at a rate 3 °C/min. Such two-stage heating significantly limited the appearance of cracks and delaminations were not observed.

Two simulated body fluids were used. Ringer's solution was prepared based on commercial tablets (Ringer's tablet, Merck, Germany) and artificial saliva according to the composition shown in Table 1. Hydrochloric acid (4M) (HCl) was used to prepare solutions of the appropriate pH values of 3, 5, and 7. The Elmetron CPI-505 pH meter was used to measure the pH values.

Table 1. Chemical composition of the artificial saliva.

Compound	Content (g/L)
$(NH_2)_2CO$	0.13
NaCl	0.7
$NaHCO_3$	1.5
Na_3HPO_4	0.26
K_2HPO_4	0.2
KSCN	0.33
KCl	1.2

The samples were cleaned in an ultrasonic washer and immersed in the prepared solutions for 3 and 12 months. After this time the test solution samples were analyzed for total titanium content at the Centre of Biological and Chemical Sciences, the University of Warsaw (Warsaw, Poland). The inductively coupled plasma mass spectrometry (ICP-MS, Perkin Elmer, Inc., Waltham, MA, USA) was carried out by the norm E2371-13 to determine the titanium content in the solutions. The ions were separated using a special mass analyzer, distributing the ions according to the value of their mass-to-charge ratio. The ICP-MS was calibrated using an external calibration curve, which was prepared using 1% nitric acid and a titanium pattern. The quadrupole mass spectrometer, Elan 9000 Perkin Elmer ICP-MS, with conjugated plasma induction excitation was used for the study.

The solutions were mineralized before measurements in a closed microwave system. Approximately 1 g of the solution and 1 mL of 30% nitric acid were used for mineralization. The samples were then diluted to 15 mL.

2.3. Calculations of Diffusion of Titanium Ions through Oxide Coatings

The last stage was the calculations of the theoretical diffusion rate of titanium atoms through the rutile crystalline structure. The calculations were based on Arrhenius and Fick's laws [38] and the earlier high-temperature measurements [39,40].

3. Results

3.1. Examinations of Implants

Figure 2 presents the surfaces of different three new implants, produced by two companies, A and B. The layer discontinuities, material defects appearing in the layer, unevenness, numerous rolling scratches, and material allowances resulting from the surface treatment processes are visible before implantation.

The used implants were obtained from surgeons from the Warsaw Medical University. Only removed implants, among all, which demonstrated clear signs of damage, were selected. When examinations of the used implants (Figure 3) were made, two areas of surfaces could be distinguished. The first was the top of the threads, with characteristic flattened, rubbed bumps, with clearly visible pits on the surface. The second type was the bottom of the threads, in which there is a detachment of the material of the layer from the ground, and numerous deep cracks are visible likely arisen as a result of stress concentration, which are potential places of corrosion initiation and development of corrosion processes in the environment of body fluids and particularly aggressive saliva.

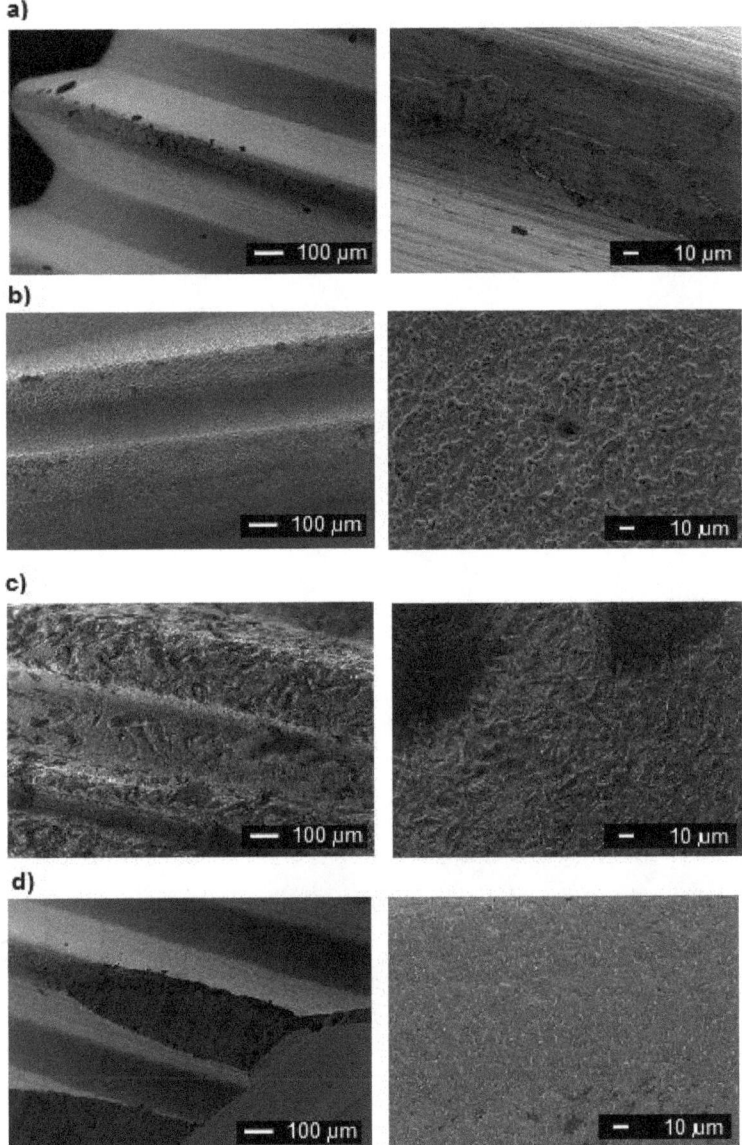

Figure 2. Surfaces of new dental implants: (**a**) A company; (**b**) B company; (**c**) C company; (**d**) D company. SEM.

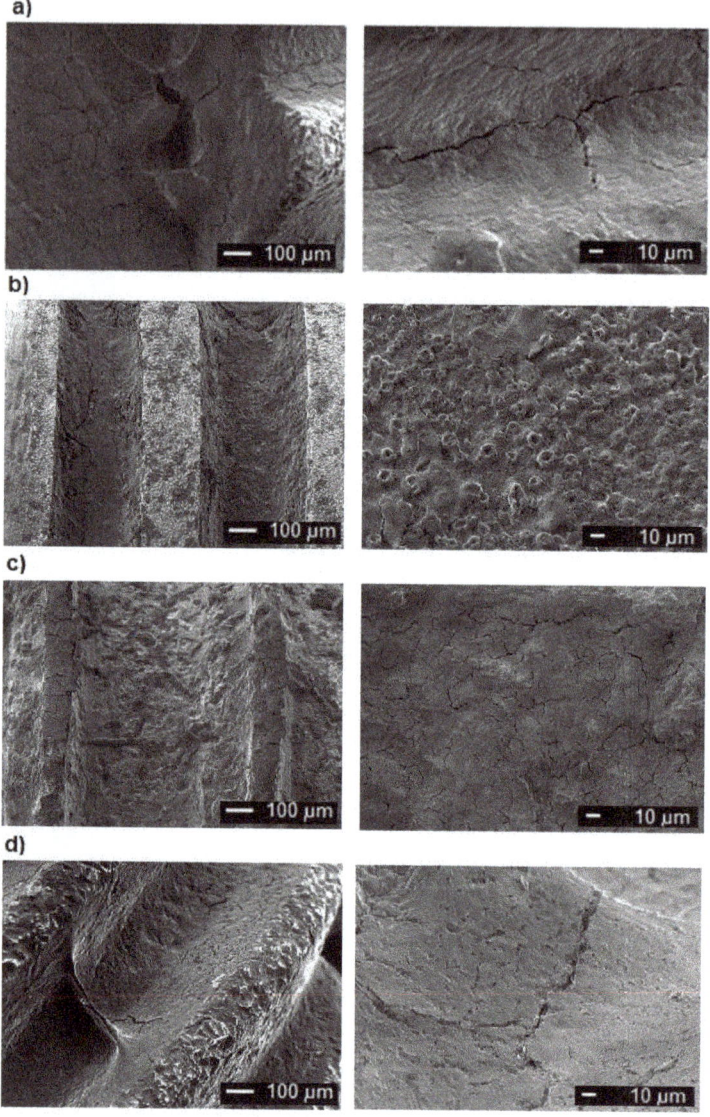

Figure 3. Surfaces of used dental implants: (**a**) A company; (**b**) B company; (**c**) C company; (**d**) D company. SEM.

The surfaces of the used implants were subjected to purification or sterilization processes, thanks to which we can observe traces of organic residues on the surface—bacteria and tissues. Surprisingly, the largest clusters of bacterial colonies are located between the tip and the bottom of the thread, on the lateral surfaces. Perhaps this phenomenon is caused by adverse conditions at the tops and bottoms of the threads. Pits are visible on the surface of the thread tops, numerous and deep cracks in the thread cavities, which may indicate a significant impact of the environment and continuous operation of the implant—the influence of tensile forces and friction forces that affect the implant placed in the bone. On the implant surfaces, discontinuities of bone formation (Figure 4a,c) and bacterial colony residues (Figure 4b,d) can be observed, which tightly cover the material.

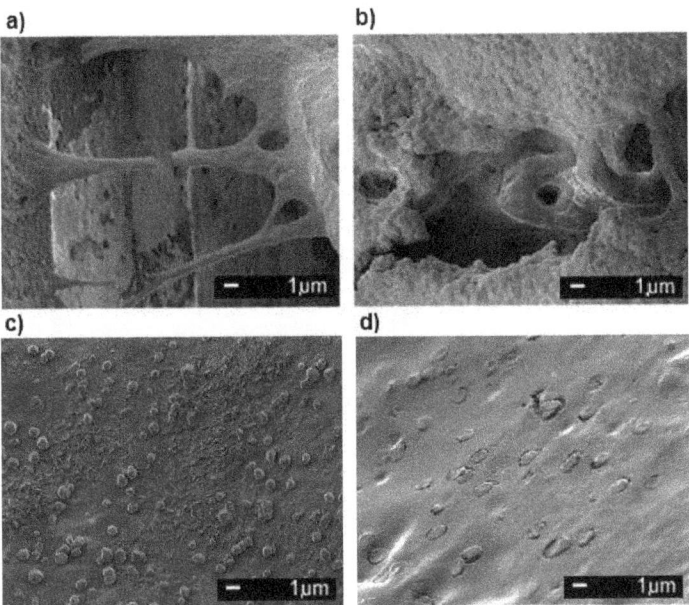

Figure 4. Organic remains on the surfaces of used dental implants: (**a**) A company; (**b**) B company; (**c**) C company; (**d**) D company. SEM.

The implant cross-sections (Figure 5) illustrate surface unevenness, numerous discontinuities characterizing the layers produced, material stratification, a significant number of cracks of varying lengths, and arrangement occurring in the coating. They can be ideal for corrosion progress.

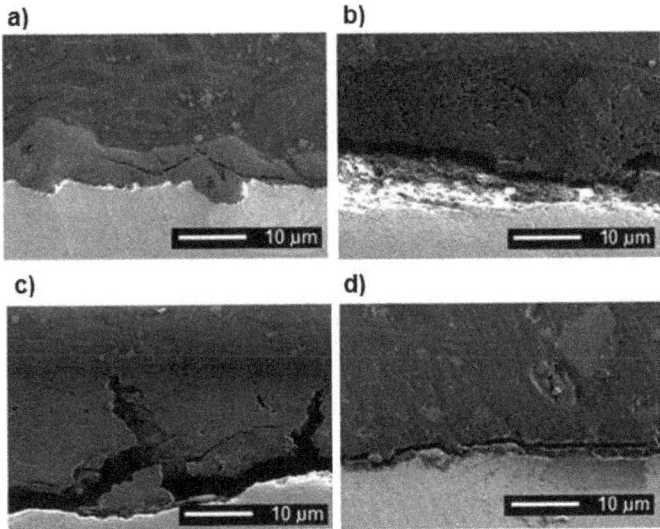

Figure 5. Cross-sections of used dental implants: (**a**) A company; (**b**) B company; (**c**) C company; (**d**) D company. SEM.

Using computed tomography, the layer thickness distribution on dental implants was depicted (Figure 6). The analysis shows that the thickness of the layers formed on the surface of the implants is diverse and does not evenly distribute. The surfaces of the vertices and thread cavities are characterized by a larger thickness of the coating. The images of the layer thickness distributions on the surface of the dental implants were very similar.

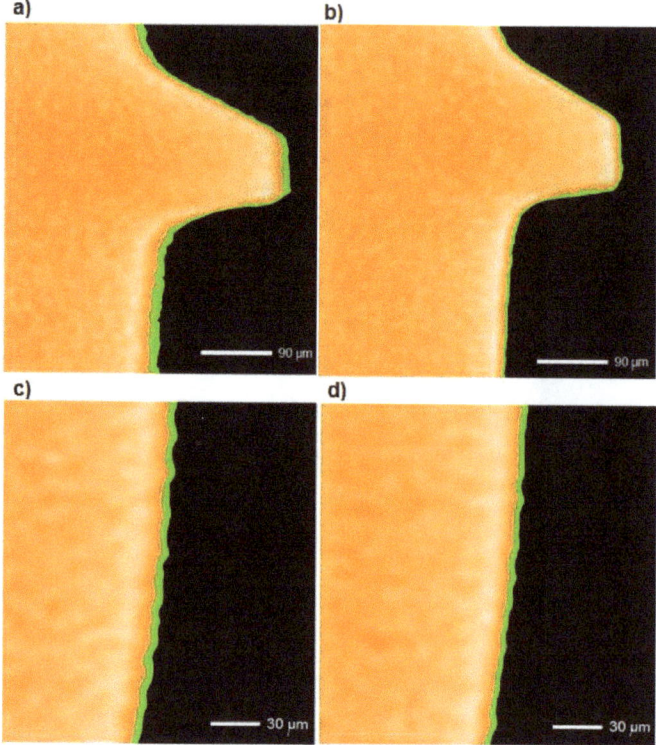

Figure 6. Images obtained by the computed tomography study of the sections: —longitudinal: (**a**) D company; (**b**) A company; —transverse dental implants: (**c**) D company; (**d**) A company.

3.2. The Dissolution of Rutile

Knowing the characteristics and defects occurring in the layers covering dental implants, in the second stage of research, the determination of the rate of penetration of titanium ions into the solution was undertaken. Titanium oxide powder samples were prepared using powder metallurgy processes, which were immersed in Ringer's solution and artificial saliva for a period of 3 and 12 months.

The surface of the samples produced, featuring a slight degree of porosity, is shown in Figure 7. The samples vary in grain size from 1.429 to 8.184 µm.

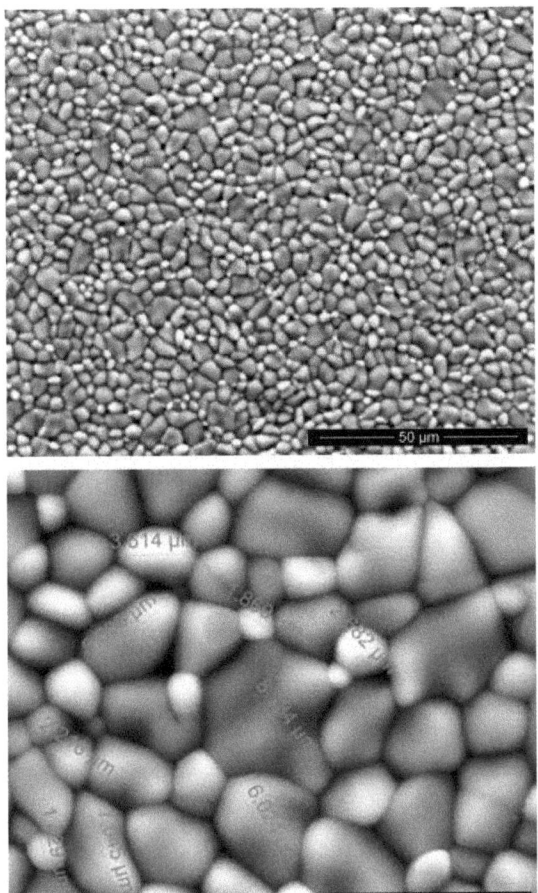

Figure 7. SEM microstructures of the sample surface after pressing with a force of 2 kN.

The analyses of the titanium ions carried out three months after their exposure to both SBFs at pH 3, 5, and 7 (Table 2), showed negligible solubility of the titanium dioxide, below the 0.080 mg/kg limit of determination. In solution adjusted to pH 3, titanium dissolution was distinctly higher, but only after 12 months, 0.093 mg/kg.

Table 2. Titanium content according to ICP-MS analysis at different pH (mg/kg).

Time	pH	Ringer's Solution	Artificial Saliva
3 months	3	<GO *	<GO *
	5	<GO *	<GO *
	7	<GO *	<GO *
12 months	3	0.093	0.136
	5	<GO *	0.107
	7	<GO *	0.107

* GO—Limit of identification of titanium 0.080 mg/kg.

Slightly different results were achieved for the artificial saliva solution. The contents of titanium ions for artificial saliva of different pH were higher than those for Ringer's solution and about 0.107 mg/kg for pH 5 and 7, while for pH 3 the titanium ion level exceeded 0.137 mg/kg.

These experiments show the importance of the pH value of body fluids and the possible dissolution and penetration of titanium ions into the human body. The increasing dissolutions of ferrous oxides [41] and copper oxides [42] with decreasing pH, and ruthenium–titanium oxide coating at pH 2 [43] are in accordance with obtained results showing the important effect at the lowest pH value. The more distinct dissolution of rutile in artificial saliva is likely due to the higher chloride concentration in saliva than in blood, and the susceptibility of the oxides to the pitting. It is worth noting that the solutions used only simulated natural human body fluids and did not contain various biological substances such as enzymes, which can create an even more aggressive environment. The important conclusion, however, is that a titanium dental implant in the mouth is more susceptible to dissolution than in other tissues. The drastic lowering of the pH value of the solution accelerates the process of removal of titanium ions from rutile, always present on the titanium surface. Thus, changes in the pH value, occurring during some inflammatory reactions in the living organism, may significantly influence the condition of the oxide layer, and consequently the status of the implant.

In these long-term, expensive tests, we have analyzed the trends: the effect of decreasing pH and test solution on the dissolution rate. Taking into account the extremely low values of dissolubility, the precision of the spectrometric measurements, and the use of slightly porous materials, we have concluded that a significant number of specimens should be applied to obtain the homogenous sample and low standard deviations. Taking this into account, our purpose has been only to recognize at least the row of the magnitude of dissolubility and how the pH effects what has been reached. The results clearly show that the rutile ceramics dissolubility is extremely low and it decreases with decreasing pH, as it is for metallic substances.

3.3. Diffusion of Titanium Ions in The Oxide Layer

To calculate the distance of the diffusion of titanium ions in rutile lattice, data of two references were taken into account. In [40], the random tetravalent titanium atoms were assumed to be the predominant defects evident from self-diffusion. The enthalpy of motion was determined as ΔH_m = 57.03% ± 4.9% kcal/mole. In another report [39], for diffusing the radio-isotope titanium-44 into single crystal rutile at temperatures in the range of 900 to 1300 °C, the activation energy was found to be 61,400 calories per mole and the frequency factor was calculated to be 6.4×10^{-4} m^2/s. Assuming the diffusion enthalpy at 59.2 kcal/mol and D_0 at the above value, the titanium diffusion coefficient at room temperature (293 K) was calculated at about 10^{-49} m^2/s. That following, the diffusion distance at this temperature in one year is about 3×10^{-34} nm.

4. Discussion

All new implants were made of the Ti6Al4V alloy by casting and milling (likely CNC). As a rule, such implants are assumed to have a perfect surface, at designed roughness achieved by mechanical treatment or chemical acidic (SLA implants) or alkaline treatment. Some of the commercial implants have deposited coatings (Osseotite and Nanotite implants). The detailed surface treatment is not disclosed. The majority of dental implants are likely subjected to micro-arc oxidation in phosphate solutions.

So far, the imperfections visible on new implants are attributed to the forces acting during implantation surgery. For example, the grooves and abraded facets, and loose titanium particles at the interface were reported for dental implants and attributed to the surgical procedure [44].

It is a damage that certainly locally destroys the titanium coatings. However, our investigations showed that several implants before any implantation possessed already imperfect surfaces with such damage forms of the oxide coating as the layer discontinuities (holes/pits), large unevenness, and rolling scratches. Such defects may initiate the local degradation of an implant, in particular the cracking and pitting corrosion.

The examinations of implants used for a relatively short time confirmed the above assumption, even indirectly. The pits are visible on the tops of the threads, and the detachment of the coating on the bottoms. The numerous deep cracks are likely arisen as a result of stress concentration, and can serve as potential places of corrosion initiation and development in the environment of body fluids and particularly aggressive saliva.

The traces of organic residues on the surface were between the tops and the bottoms of the threads. Perhaps this phenomenon is caused by adverse conditions at the tops and bottoms of the threads due to a significant impact of the environment and continuous influence of tensile and friction forces that affect the implant placed in the bone. The detailed mechanism of this phenomenon cannot, however, be proposed at the moment.

Three possible mechanisms of the release of titanium ions can be proposed as already shown. These results demonstrate that the most significant is corrosion initiation and propagation of corrosion in a presence of local damage of oxide coating and, on the other hand, complex stresses imposed on the screw implants. However, it is necessary to consider whether two other mechanisms can also operate and be comparable.

The dissolution of rutile may occur, but at an extremely low rate and only in strongly acidic environments. Such conditions may occur only in inflammation conditions at which pH may reach highly acidic values. Even if so, the dissolution rate achieves 0.136 mg/kg in 12 months, such results means that the oxide coating even 100 μm thick (after anodic oxidation) decreases less than 1 nm. Such a mechanism is then impossible and must be rejected.

The third mechanism, which can be considered, is diffusion of titanium ions through the rutile lattice. The diffusion of titanium at the temperature of a human body in the rutile crystalline structure seems unlikely. There is no such data even for high temperatures so that it seems desirable to consider the titanium diffusion in other structures. The performed calculations showed the titanium diffusion coefficient as extremely low, even below that for the diffusion of titanium in yttria-stabilized zirconia, the diffusion coefficient at room temperature is below 10^{-30} m^2/s [45]. It means that the time necessary to diffuse through a 10 nm thick oxide layer would be as high as 10^{35} s.

Summarizing, it can be said that the only origin of the degradation processes resulting in, among other causes, in a necessity of removal of the dental implant, is the damage of the oxide coating. Such degradation may be attributed to the forces during implantation surgery, but they are likely initiated by the cracks, crevices, and discontinuities already appearing at the manufacturing stage.

5. Conclusions

The titanium dissolution occurs only by the corrosion tunnels in the oxide layer. The tunnels may be formed by cracks or discontinuities. Such potential corrosion initiation and development sites are already present in new implants and they become operative in applied implants during their use.

The present results demonstrate that among three possible mechanisms such as (i) diffusion of the liquid environment into the cracks and crevices in oxide coating; (ii) chemical dissolution of the titanium oxide layer; and (iii) diffusion of titanium atoms through the oxide layer, the two last processes are very unlikely to cause the damage of dental implants.

The both dissolution of rutile and titanium diffusion through the perfect oxide structure are negligible at the temperature of the human body. However, when the pH value at the implant surface and in the environment of saliva falls locally, the oxide layer starts to dissolve, but even at pH = 3, only a small fraction, 10^{-8} of the rutile oxide, may dissolve during 12 months.

Author Contributions: Methodology, A.O.; validation, A.O.; investigation, A.O.; resources, A.O.; original draft preparation, A.O.; formal analysis, A.Z.; writing—conceptualization, A.O. and A.Z.; writing—review & editing, A.O. and A.Z. All authors have read and agreed to the published version of the manuscript.

Funding: This research received no external funding.

Acknowledgments: We are grateful to Grzegorz Gajowiec (GUT) for his examinations of oxidized surfaces with the SEM, M.Eng. Gabriel Strugała for his examinations of oxidized surfaces with the CT, Eliza Kurek from Biological and Chemical Research Center University of Warsaw for the ICP-MS studies, and Andrzej Wojtowicz from the Department of Dental Surgery, the Medical University of Warsaw for delivery of the implants.

Conflicts of Interest: The authors declare no conflict of interest.

References

1. Huang, H.H. Effects of fluoride concentration and elastic tensile strain on the corrosion resistance of commercially pure titanium. *Biomaterials* **2002**, *23*, 59–63. [CrossRef]
2. Toumelin-Chemla, F.; Rouellet, F.; Burdairon, G. Corrosive properties of fluoride containing odontologic gels against titanium. *J. Dent.* **1996**, *24*, 109–115. [CrossRef]
3. Vargas, E.; Baier, R.; Meyer, A. Reduced corrosion of cp Ti and Ti–6Al–4V alloy endosseous dental implant after glow discharge treatment: A preliminary report. *Int. J. Oral Maxillofac. Implants* **1992**, *7*, 338–344. [PubMed]
4. Agarwal, A.; Tyagi, A.; Ahuja, A.; Kumar, N.; De, N.; Bhutani, H. Corrosion aspect of dental implants—An overview and literature review. *Open J. Stomatol.* **2014**, *4*, 56–60. [CrossRef]
5. Lavos-Valereto, I.C.; Wolynec, S.; Ramires, I.; Guastaldi, A.C.; Costa, I. Electrochemical impedance spectroscopy characterization of passive film formed on implant Ti–6Al–7Nb alloy in Hank's solution. *J. Mater. Sci. Mater. Med.* **2004**, *15*, 55–59. [CrossRef]
6. González, J.E.G.; Mirza-Rosca, J.C. Study of the corrosion behavior of titanium alloys for biomedical and dental implants applications. *J. Electroanal. Chem.* **1999**, *471*, 109–115. [CrossRef]
7. Cigada, A.; Cabrini, M.; Pedeferri, P. Increasing of the corrosion resistance of the Ti6Al4V alloy by high thickness anodic oxidation. *J. Mat. Sci. Mat. Med.* **1992**, *3*, 408–412. [CrossRef]
8. Fathi, M.H.; Salehi, M.; Saatchi, A.; Mortazavi, V.; Moosavi, S.B. In vitro corrosion behavior of bioceramic, metallic, and bioceramic-metallic coated stainless steel dental implants. *Dent. Mat.* **2003**, *19*, 188–198. [CrossRef]
9. Kasemo, B.; Lausmaa, J. Biomaterial and implant surfaces: A surface science approach. *Int. J. Oral Maxillofac. Implants* **1988**, *3*, 247–259.
10. Aziz-Kerrzo, M.; Conroy, R.G.; Fenelon, A.M.; Farrell, S.T.; Breslin, C.B. Electrochemical studies on the stability and corrosion resistance of titanium-based implants materials. *Biomaterials* **2001**, *22*, 1531–1539. [CrossRef]
11. Hanawa, T. Metals ions released from metal implants. *Mat. Sci. Eng. C* **2004**, *24*, 745–752. [CrossRef]
12. Uo, M.; Watari, F.; Yokoyama, A.; Matsuno, H.; Kawasaki, T. Visualization and detectability of rarely contained elements in soft tissue by X-ray scanning analytical microscopy and electron probe micro analysis. *Biomaterials* **2001**, *22*, 1787–1794. [CrossRef]
13. Przybyszewska-Doroś, I.; Okrój, W.; Walkowiak, B. Surface modifications of metallic implants. *Eng. Biomater.* **2005**, *43–44*, 52–62.
14. De Sena, L.A.; Rocha, N.C.C.; Andrade, M.C.; Soares, G.A. Bioactivity assessment of titanium sheets electrochemically coated with thick oxide film. *Surf. Coat. Technol.* **2003**, *166*, 254–258. [CrossRef]
15. Lavos-Valereto, I.C.; König, B.; Rossa, C., Jr.; Marcantonio, E.; Zavaglia, A.C. A study of historical responses from Ti–6Al–7Nb alloy dental implants with and without plasma-sprayed hydroxyapatite coatings in dogs. *J. Mat. Sci. Mat. Med.* **2001**, *12*, 273–276. [CrossRef] [PubMed]
16. Yang, B.; Uchida, M.; Kim, H.-M.; Zhang, X.; Kokubo, T. Preparation of bioactive titanium metal via anodic oxidation treatment. *Biomaterials* **2004**, *25*, 1003–1010. [CrossRef]
17. Zieliński, A.; Sobieszczyk, S. Corrosion of titanium biomaterials, mechanisms, effects and modelisation. *Corros. Rev.* **2008**, *26*, 1–22. [CrossRef]
18. Long, M.; Rack, H.J. Titanium alloys in total joint replacement—A materials science perspective. *Biomaterials* **1998**, *19*, 1621–1639. [CrossRef]
19. Veys-Renaux, D.; Ait El Haj, Z.; Rocca, E. Corrosion resistance in artificial saliva of titanium anodized by plasma electrolytic oxidation in Na_3PO_4. *Surf. Coat. Techn.* **2016**, *285*, 214–219. [CrossRef]

20. Revathi, A.; Borrás, A.D.; Muñoz, A.I.; Richard, C.; Manivasagam, G. Degradation mechanisms and future challenges of titanium and its alloys for dental implant applications in oral environment. *Mat. Sci. Eng. C* **2017**, *76*, 1354–1368. [CrossRef]
21. Strietzel, R.; Hösch, A.; Kalbfleisch, H.; Buch, D. In vitro corrosion of titanium. *Biomaterials* **1998**, *19*, 1495–1499. [CrossRef]
22. Koike, M.; Fujii, H. The corrosion resistance of pure titanium in organic acids. *Biomaterials* **2001**, *22*, 2931–2936. [CrossRef]
23. Wierzchoń, T.; Czarnowska, E.; Krupa, D. *Surface Engineering in the Production of Titanium Biomaterials*; Printing House PW: Warsaw, Poland, 2004.
24. Haynes, D.R.; Rogers, S.D.; Hay, S. The differences in toxicity and release of bone-resorbing mediators induced by titanium and cobalt-chromium-alloy wear particles. *J. Bone Jt. Surg. Am.* **1993**, *75*, 825–834. [CrossRef]
25. Wu, H.H.; Trinkle, D.R. Direct diffusion through interpenetrating networks: Oxygen in titanium. *Phys. Rev. Lett.* **2011**, *107*, 045504. [CrossRef]
26. Sefer, B. Oxidation and Alpha-Case Phenomena in Titanium Alloys Used in Aerospace Industry: Ti-6Al-2Sn-4Zr-2Mo and Ti-6Al-4V. Bachelor's Thesis, Luleå University of Technology, Luleå, Sweden, September 2014.
27. *Aerospace Series Test Method Titanium and Titanium Alloys Part 009-Determinantion of Surface Contaminantion*; Swedish Standards Institute: Stockholm, Sweden, 2007.
28. Hanawa, T.; Asami, K.; Asaoka, K. Repassivation of titanium and surface oxide film regenerated in simulated bioliquid. *J. Biomed. Mat. Res.* **1998**, *40*, 530–538. [CrossRef]
29. Grant, D.M.; Lo, W.J.; Parker, K.H.; Parker, T.L. Biocompatible and mechanical properties of low temperature deposited quaternary (Ti, Al, V) N coatings on Ti6Al4V titanium alloy substrates. *J. Mat. Sci. Mat. Med.* **1996**, *7*, 576–584. [CrossRef]
30. Li, J.; Jansen, J.A.; Walboomers, X.F.; van den Beucken, J.J.P. Mechanical aspects of dental implants and osseointegration: A narrative review. *J. Mech. Behav. Biomed. Mat.* **2020**, *103*, 103574. [CrossRef]
31. Cheung, K.H.; Pabbruwe, M.B.; Chen, W.-F.; Koshy, P.; Sorrell, C.C. Effects of substrate preparation on TiO_2 morphology and topography during anodization of biomedical Ti6Al4V. *Mat. Chem. Phys.* **2020**, *252*, 123224. [CrossRef]
32. Kurup, A.; Dhatrak, P.; Khasnis, N. Surface modification techniques of titanium and titanium alloys for biomedical dental applications: A review. *Mater. Today Proc.* **2020**. [CrossRef]
33. Alves, A.C.; Wenger, F.; Ponthiaux, P.; Celis, J.-P.; Pinto, A.M.; Rocha, L.A.; Fernandes, J.C.S. Corrosion mechanisms in titanium oxide-based films produced by anodic treatment. *Electrochim. Acta* **2017**, *234*, 16–27. [CrossRef]
34. Cui, W.F.; Jin, L.; Zhou, L. Surface characteristics and electrochemical corrosion behavior of a pre-anodized microarc oxidation coating on titanium alloy. *Mater. Sci. Eng. C* **2013**, *33*, 3775–3779. [CrossRef]
35. Shokouhfar, M.; Allahkaram, S.R. Effect of incorporation of nanoparticles with different composition on wear and corrosion behavior of ceramic coatings developed on pure titanium by micro arc oxidation. *Surf. Coat. Techn.* **2017**, *309*, 767–778. [CrossRef]
36. Sharma, A.; Mcquillan, A.J.; Sharma, L.A.; Waddell, J.; Shibata, Y.; Duncan, W. Spark anodization of titanium–zirconium alloy: Surface characterization and 216 bioactivity assessment. *J. Mater. Sci. Mater. Med.* **2015**, *26*, 1–11. [CrossRef]
37. Ossowska, A.; Zieliński, A.; Supernak, M. Electrochemical oxidation and corrosion resistance of the Ti13Nb13Zr alloy. *Eng. Biomat.* **2013**, *XVI*, 4–6.
38. Callister, W.D. *Materials Science and Engineering—An Introduction*; Wiley: New York, NY, USA, 2007.
39. Venkatu, D.A.; Poteat, L.E. Diffusion of titanium of single crystal rutile. *Mater. Sci. Eng.* **1970**, *5*, 258–262. [CrossRef]
40. Akse, J.R.; Whitehurst, H.B. Diffusion of titanium in slightly reduced rutile. *J. Phys. Chem. Solids* **1978**, *39*, 457–465. [CrossRef]
41. Schwertmann, U. Solubility and dissolution of iron oxides. *Plant Soil* **1991**, *130*, 1–25. [CrossRef]
42. Ko, C.K.; Lee, W.G. Effects of pH variation in aqueous solutions on dissolution of copper oxide. *Surf. Interface Anal.* **2010**, *42*, 1128–1130. [CrossRef]

43. Uzbekov, A.A.; Klement'eva, V.S. Radiochemical investigation of the selectivity of dissolution of components of ruthenium-titanium oxide anodes (ORTA) in chloride solutions. *Sov. Electrochem.* **1985**, *21*, 758–763.
44. Streckbein, P.; Wilbrand, J.-F.; Kähling, C.; Pons-Kühnemann, J.; Rehmann, P.; Wöstmann, B.; Howaldt, H.-P.; Möhlhenrich, S.C. Evaluation of the surface damage of dental implants caused by different surgical protocols: An in vitro study. *Int. J. Oral Maxillofac. Surg.* **2019**, *48*, 971–981. [CrossRef]
45. Kowalski, K.; Bernasik, A.; Sadowski, A. Bulk and grain boundary diffusion of titanium in yttria-stabilized zirconia. *J. Eur. Ceram. Soc.* **2000**, *20*, 951–958. [CrossRef]

© 2020 by the authors. Licensee MDPI, Basel, Switzerland. This article is an open access article distributed under the terms and conditions of the Creative Commons Attribution (CC BY) license (http://creativecommons.org/licenses/by/4.0/).

Article

Influence of Two-Stage Anodization on Properties of the Oxide Coatings on the Ti–13Nb–13Zr Alloy

Agnieszka Ossowska [1,*], Andrzej Zieliński [1], Jean-Marc Olive [2], Andrzej Wojtowicz [3] and Piotr Szweda [4]

1. Department of Materials Engineering and Bonding, Gdańsk University of Technology, 80233 Gdańsk, Poland; andrzej.zielinski@pg.edu.pl
2. Institut de Mécanique et d'Ingénierie, Université de Bordeaux, 33405 Talence CEDEX, France; jean-marc.olive@u-bordeaux.fr
3. Department of Dental Surgery, Medical University of Warsaw, 02097 Warsaw, Poland; awojt@kcs.amwaw.edu.pl
4. Department of Pharmaceutical Technology and Biochemistry, Gdańsk University of Technology, 80233 Gdańsk, Poland; piotr.szweda@pg.edu.pl
* Correspondence: agnieszka.ossowska@pg.edu.pl; Tel.: +48-58-347-19-63

Received: 2 June 2020; Accepted: 20 July 2020; Published: 22 July 2020

Abstract: The increasing demand for titanium and its alloys used for implants results in the need for innovative surface treatments that may both increase corrosion resistance and biocompatibility and demonstrate antibacterial protection at no cytotoxicity. The purpose of this research was to characterize the effect of two-stage anodization—performed for 30 min in phosphoric acid—in the presence of hydrofluoric acid in the second stage. Scanning electron microscopy, atomic force microscopy, energy-dispersive X-ray spectroscopy, X-ray diffraction, Raman spectroscopy, glow discharge optical emission spectroscopy, nanoindentation and nano-scratch tests, potentiodynamic corrosion studies, and water contact angle measurements were performed to characterize microstructure, mechanical, chemical and physical properties. The biologic examinations were carried out to determine the cytotoxicity and antibacterial effects of oxide coatings. The research results demonstrate that two-stage oxidation affects several features and, in particular, improves mechanical and chemical behavior. The processes influencing the formation and properties of the oxide coating are discussed.

Keywords: titanium alloys; electrochemical oxidation; nanotubular oxide layers; microstructure; nanomechanical properties; corrosion resistance; wettability; antibacterial protection; cytotoxicity

---

1. Introduction

Titanium and its alloys—due to their mechanical properties—excellent corrosion resistance, and a high strength/density ratio, are nowadays the most appropriate materials for load-bearing implants and biomedical materials [1,2] used, e.g., in arthroplasty [2,3], as dental implants [4–6] and dental prostheses [7]. The titanium and its alloys proposed for medicine, after their oxidation, include medical titanium [8–10], Ti–6Al–4V [11,12], Ti–6Al–7Nb and Ti–13Nb–13Zr [13] alloys. The most commonly used Ti–6Al–4V alloy contains alloying elements, which may provoke undesirable tissue reactions damaging nerves cells, softening the bones, and, as a consequence, resulting in the appearance of diseases of the circulatory and central nervous systems [14,15]. Therefore, Ti-13Zr-13Nb alloy was chosen for this research as it has no harmful elements and possesses a low (76 GPa) Young's modulus, similar to that of cortical bone, providing better stress distribution at the implant–bone contact zone and preventing against loosening and damage of the implant.

The different surface modification methods of titanium alloys such as deposition of coatings, oxidation, ion beam surface modification, ion implantation, titanium plasma spraying, acid etching, grits blasting, sandblasting followed by acidic etching, electropolishing and laser melting were applied for titanium and its alloys for biomedical applications [16]. In particular, adhesion of the cells has been shown better on rough than on smooth surface [17,18]. Such surface characteristics may be achieved by the laser treatment [19,20], surface mechanical attrition [21,22], acid etching [23,24], deposition of phosphate [25–29] and composite coatings [30–32]. Among those modifying approaches, the oxidation remains essential as it can either form nanotubular oxide structures or rough oxide surfaces, enhancing the adhesion of osteoblasts, if alone, and deposition of coatings, if used as an interlayer.

Oxidation plays an essential role among possible surface engineering methods, and even a spontaneously formed titanium oxide layer is a barrier limiting the entry of metal ions into tissues [33]. Moreover, the oxide layer may influence the osteoinduction processes by a change in the architectural features and chemical composition of the oxides [34]. The techniques used for this purpose include the low voltage anodization, micro-arc oxidation (MAO) [35], thermal oxidation of titanium biomaterials [36], less often the oxidation using hydrogen peroxide [37] and laser-enhanced oxidation [38].

The MAO is used as a technique for creating a multiporous or highly developed surface, often implemented with different ions [39–42]. A novel "cortex-like" micro/nano dual-scale structured TiO_2 coating was prepared in such a way in tetraborate electrolytes [43]. The MAO in ammonium acetate was resulted in a multiporous, crystalline titanium oxide layer demonstrating the apatite forming ability [9]. The antibacterial activity may be achieved by the MAO performed in electrolytes comprising Ag, Cu or Zn [40]. However, this technique needs a high voltage and results in thick oxide coatings.

The low potential electrochemical oxidation may result in either the compact oxide [44] or nanotubular oxide coatings, depending on the type of electrolyte and anodization parameters [45–55]. The TiO_2 nanotubular surface provided topography favorable for improving the clinical performance of implants when comparing to the sand-blasted acid-etched topography [56]. The individual nanotubes can be filled with antibiotics or nanometals for introducing the antibacterial ability. The release rate of nanosilver depends on its placement: relatively fast release was observed for nanoAg inside the nanotubes and gradual release, for Ag inside the cavities [57]. The functionalization of titanium dioxide nanotubes with some biomolecules was developed for biomedical applications [58] and the osteogenic differentiation can be modulated by various additional treatments of nanotube coatings on Ti–6Al–4V implants [12]. Superhydrophobic titanium oxide nanotube arrays may serve as the drug reservoir, and ultrasonic waves may trigger the drug release [59]. Such a superhydrophobic Ti surface was fabricated by subsequent anodization in H_2O_2 followed by aging [60]. Hierarchical structures were obtained, applying two nanotexturing surface treatments onto titanium coatings, anodic oxidation and alkaline treatments, and the simultaneous presence of micro-/nano-roughness resulted in a distinct increase in cell proliferation [61].

Different composite coatings were also developed. The ion implantation of helium ions was made on the oxide film obtained by previous anodization to improve hydrophilic properties [8]. The decoration of previous titanium oxide nanotubes with MnO increased the ability to form apatite [62]. Osseointegration was enhanced by coating the titanium implants with a nanostructured thin film comprised of titanium carbide and titanium oxides clustered around graphitic carbon [63].

The oxides obtained by low voltage anodization can be in the form of thin, compact coatings or nanotubular layers. In the past, the bi-layer coating was prepared [47] by gas oxidation of titanium alloy and then electrochemical oxidation resulting in nanotubular layers grown on the previous compact oxide layer. Such treatment brought out the highly corrosion-resistant coatings but possessing the relatively short nanotubes. Therefore, the present research has applied the two-stage electrochemical oxidation assuming that such procedure may positively affect some properties of the oxide coatings, in which the nanotubular layer is formed not in the bare metal, but in the compact oxide layer. In particular, the mechanical and chemical behavior have been expected to improve.

2. Materials and Methods

The study was performed on a two-phase titanium alloy Ti–13Nb–13Zr of chemical composition listed in Table 1. The microstructure of the investigated alloy (Baoji SeaBird Metal Materials Co., Ltd., Baoji, China) is shown in Figure 1. It is a β-phase structure comprising of the α' phase being a supersaturated solution with a slightly stubborn effect and a martensitic structure, which is formed as a result of rapid cooling from the temperature of the β-phase stability or as a result of plastic deformation.

Table 1. Chemical composition of the Ti–13Nb–13Zr alloy, weight percent (according to manufacturer's certificate).

Nb	Zr	Fe	C	N	H	O	Ti
13.5	13.5	0.05	0.04	0.013	0.004	0.11	bal.

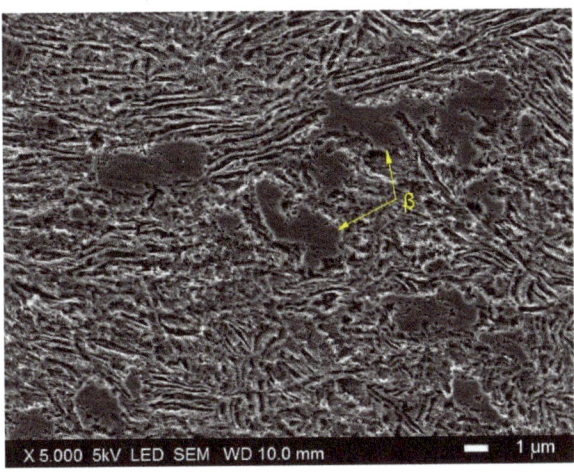

Figure 1. Microstructure of the Ti–13Nb–13Zr alloy after etching with the Kroll solution composed of 2 mL HF (40 wt.%), 2 mL HNO_3 (55 wt.%) + 96 mL H_2O.

The specimens of dimensions 15 mm × 10 mm × 4 mm were cut from the alloy sheet of initial thickness of 4.2 mm. Then the samples were ground with abrasive papers (No. 2500 as the last, Struers, Inc., Cleveland, OH, USA). Afterward, the specimens were cleaned in an ultrasonic chamber (Sonic-2, Polsonic Palczynski Sp. J., Warsaw, Poland) with isopropanol, methanol (Avantor Performance Materials Poland S.A., Gliwice, Poland) and distilled water, subsequently, for 5 min in each batch and finally dried in cold air.

The tests were performed in a standard circuit composed of an electrochemical cell, power supply (SPN-110-1C, MPC Lab Electronics, Nijmegen, The Netherlands), Pt electrode as the polarizing electrode and the tested metallic electrode. Neither stirring, aeration nor deaeration were applied. All measurements were performed at room temperature. The anodization parameters were set up based on some earlier investigations [47]; in particular, even if the electrochemical oxidation time has a small effect on the oxide thickness, the 30 min period was assumed necessary to perform the electrochemical oxidation at equilibrium conditions.

The anodization was carried out at the first stage electrochemically in 1 M orthophosphoric acid (H_3PO_4) at the potential value of 40 V, in one step, at 20 °C, for 30 min (samples obtained in such a way are here designated as EO1). The electrochemical oxidation was repeated in 1 M orthophosphoric acid with an addition of 0.3 vol.% of hydrofluoric acid (HF) (designation EO2). The process was performed again at 20 °C, at a potential value of 20 V, in one step, for 30 min. The coatings were also obtained by two-stage oxidation—first EO1, then EO2 (designation EO1 + EO2). After each of the processes, the samples were rinsed in distilled water and dried in cold air. The samples were heat-treated after oxidation at 400 °C for 2 h in the air (humidity <70%).

The surfaces of specimens and their cross-sections after each form of oxidation were examined with the scanning electron microscope (SEM JEOL JSM-7600 F, JEOL, Ltd., Tokyo, Japan), equipped with a LED detector, at 5 kV acceleration voltage. The chemical composition of the coatings was determined using an X-ray energy-dispersive spectrometer (EDS, Edax, Inc., Mahwah, NJ, USA).

The surface examinations, with used linear roughness measurement, were performed with the atomic force microscopy (MFP-3D, Oxford Instruments Asylum Research Inc., Santa Barbara, CA, USA) at the Université Bordeaux, France. The surface topography was assessed in the noncontact mode at a force 50 mN. The roughness index R_a was estimated within an area of 5.0 μm × 5.0 μm.

The X-ray diffraction studies were carried out at the Gdańsk University of Technology, Faculty of Applied Physics and Mathematics, with the use of X-ray diffractometer (Philips X'Pert Pro–MPD, Brighton, UK) system with a vertical T–T goniometer (190-mm radius). The X-ray source was a long-fine-focus, ceramic X-ray tube with Cu anode. The standard operating power was 40 kV, 50 mA (2.0 kW). The system optics consisted of programmable divergence, anti-scatter and receiving slits, incident and diffracted beam Soller slits, curved graphite diffracted beam monochromator and a proportional counter detector (Bragg–Brentano parafocusing geometry (2θ ca. 5°–100°). The spectroscopic examinations of the grown oxide layers were performed with the Raman spectrometer (Horiba Jobin Yvon GmbH, Bensheim, Germany) at the Max Bergmann Centrum of Biomaterials, Dresden Technical University.

The glow discharge optical emission spectroscopy (GDOES) tests were carried out at the University of Bordeaux, using the GD-Profiler 2 (Horiba Jobin Yvon IBH Ltd., Glasgow, UK). The measurements were performed using the following process parameters: a glow discharge source (argon plasma) at 700 Pa and 30 W, measurement time 120 s.

The nanoindentation tests were performed with the NanoTest Vantage (Micro Materials, Wrexham, UK) equipment using a Berkovich three-sided pyramidal diamond. The maximum applied force was equal to 5 mN, the loading and unloading times were set at 20 s, the dwell period at full load was 10 s. The distances between the subsequent indents were 50 μm. During the indent, the load–displacement curves were determined using the Oliver and Pharr method. Based on the load–penetration curves, the surface hardness (H) and reduced Young's modulus (E) were calculated using the integrated software. The critical process parameters included the maximum force, holding time and test rate. In calculating Young's modulus (E), a Poisson's ratio of 0.3 was assumed for the titanium oxide layer. The measurements were processed in randomly selected five points for each surface, and the results were averaged.

The electrochemical measurements of corrosion parameters were performed by a potentiodynamic mode in the Ringer's solution. The simulated body fluid was obtained by dissolving a Ringer's tablet (Merck KGaA, Darmstadt, Germany; each tablet contained 1.125 g NaCl, 0.0525 g KCl, 0.03 g $CaCl_2$ and 0.025 g $NaHCO_3$) in 0.5 L of distilled water at 20 °C. Different pH levels were obtained by adding the hydrochloric acid (5 wt.% to the solution. Lowering of pH even to 3 resembled acidic environmental conditions during inflammation [64], so the test was carried for pH ranging from 7 (normal physical state) through 5 to 3 (inflammatory state). A standard three-electrode electrochemical cell was used comprising of a saturated calomel electrode (SCE) as the reference electrode, a platinum electrode as the counter electrode and the sample as the working electrode (anode). All experiments were performed using a potentiostat/galvanostat (VersaSTAT 4, Ametek Scientific Instrumentation, Leicester, UK). Before the test, the samples were stabilized at their open circuit potential (OCP) for 0.5 h. Potentiodynamic polarization tests were carried out at a potential change rate of 10 mV/min, within a scan range from −2 to 2.5 V. The corrosion potential E_{corr} and corrosion current density i_{corr} were determined from the polarization curves using the Tafel extrapolation method.

The water contact angle (wettability) measurements were taken for the reference Ti–13Nb–13Zr alloy and oxidized specimens using a contact angle goniometer (Attension Thete Lite, Dyne Technology, Lichfield, UK) at room temperature. All analyses were repeated three times for each sample.

Studies of antibacterial activity of nanotubular surfaces were carried out at the Gdańsk University of Technology, Faculty of Chemistry, with the *Staphylococcus aureus* ATCC25923 strain. The samples were put in 5 mL of bacterial suspension (containing at least 10^6 colony forming units (CFU) in 1 cm^3) prepared in phosphate buffered saline (PBS, chemical composition 8.0-g/L NaCl, 0.2-g/L KCl, 1.44-g/L Na_2HPO_4, 0.24-g/L KH_2PO_4)), in which they stayed for 1 min. This step of the procedure aimed to allow the bacterial cells to adsorb on the surfaces of the tested materials. Next, the samples (with bacteria adsorbed on their surfaces) were transferred to 5 mL of sterile TSB medium placed in the 8-well microplates. The samples were incubated at 37 °C for 24 h (one day) or 120 h (5 days). Subsequently, the samples were removed carefully from TSB medium and rinsed by submersion three times in a sterile saline solution (0.9% NaCl). Afterward, the samples were placed in the wells of a new titration plate containing 5 mL of MTT (3-(4,5-dimethyl-2-thiazolyl)-2,5-diphenyl-2*H*-tetrazolium bromide) solution (0.3%) in PBS. The living cells of bacteria reduce MTT to insoluble in water violet formazan crystals, and the amount of formed formazan is proportional to the number of live bacteria that are still present (in the form of biofilm) on the surfaces. Following 2 h incubation at 37 °C in the dark, the solution of MTT in PBS was carefully removed from the wells and replaced with 5 mL of DMSO for dissolving formed formazan crystals. The optical density of the obtained solutions was measured at 540 nm using a Victor^3 microtiter reader (PerkinElmer, Waltham, MA, USA).

Cytotoxicity tests were performed at the Warsaw Medical University, Department of Dental Surgery. They were carried out on the titanium alloy and the oxidized surfaces of the samples. Experiments were performed on fibroblasts obtained from neonatal rat Lewis Op/Op after the third passage. A small microscope slide was placed into small plastic plates with a diameter of 35 mm (430165, Corning Manufacturer, Corning, NY, USA). For all of them, except for control plates, single titanium samples were filled with a suspension of cells in the culture medium. All plates received 100,000 cells suspended in 2.0 mL medium. After five days, the slides with the cells deposited on them were rinsed with physiological saline and preserved in a mixture of methanol and acetic acid (3:1) for 5 min, then stained with hematoxylin and eosin. The preparations were dehydrated with DPX (a mix of distyrene, a plasticizer and xylene), dried and subjected to microscopic evaluation. The density of cultured cells and their morphologic features, as well as the presence of forms of mitotic divisions, were assessed. The evaluation of each sample was carried out three times.

3. Results

3.1. Microstructure, Surface Topography, Phase and Chemical Compositions

Figure 2 presents the morphology of the oxide coatings. They all were homogenous and transparent, but the interference of reflected light resulted in a color effect related to the applied voltage and resultant thickness of the oxide layer and its structure. The samples after EO1 treatment showed a blue color (Figure 2a), typical of titanium oxidized at 40 V and resulting in the thickness of about 74 nm [65]. The oxide coatings obtained after EO2 treatment and EO1 + EO2 modification were matt-gray as expected for nanotubular layers.

The observations of the surfaces of oxide coatings revealed a homogenous and even surface after EO1 treatment (Figure 2a) and an appearance of nanotubes after EO2 (Figure 2c) and EO1 + EO2 (Figure 2e) surface modifications.

The measurement of the thickness of the thin EO1 coating was challenging as before oxidation, the surface was mirror-like, and there were reflections from the surface. The results of measurements based on the cross-sections of the samples (Figure 2b,d,f) showed that the EO1 coating was about 80-nm-thick (Figure 2b), in perfect accordance with the previously cited report, the EO2 coating was about 1000-nm-thick (Figure 2d), and the EO1 + EO2 coating had similar thickness. However, the last coating could be supposed to compose of two zones: typical nanotubular outer layer and an inner layer of presumably different view as discussed later (Figure 2f).

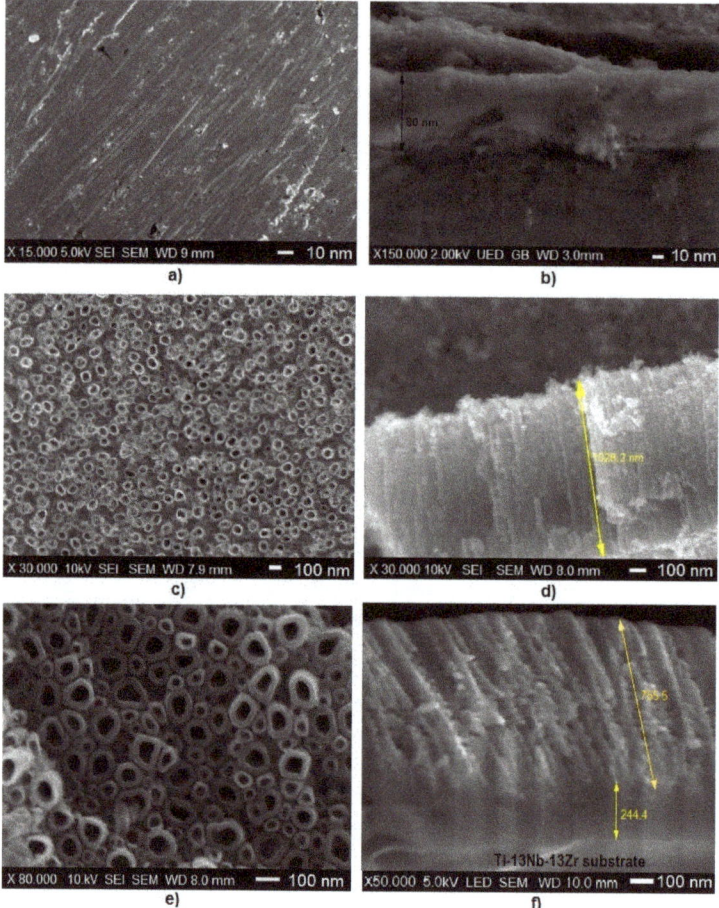

Figure 2. Images SEM - comparison of (**a**) alloy after amorphous layers (EO1) treatment (surface); (**b**) alloy after EO1 treatment (cross-section); (**c**) alloy after nanotubular layer (EO2) treatment (surface); (**d**) alloy after EO2 treatment (cross-section); (**e**) alloy after EO1 + EO2 treatment (surface), (**f**) alloy after EO1 + EO2 treatment (cross-section).

Figure 3 shows the surface coating topography after different electrochemical oxidation and Table 2 presents the roughness of coatings. After EO1 treatment, the smoothest layer, even compared to the polished material, was observed. The EO2 (not shown in figure) and EO1 + EO2 coatings were characterized by slightly increased roughness than the EO1 and substrate material.

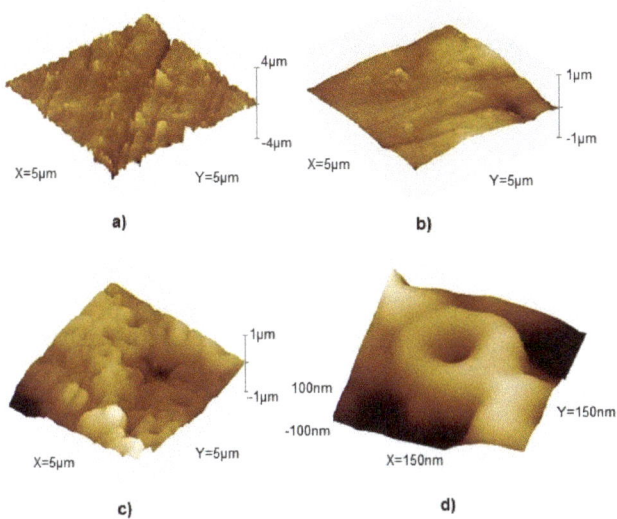

Figure 3. Atomic Force Microscope (AFM) images of the surface topography. (a) Substrate titanium alloy; (b) alloy after EO1 treatment; (c) alloy after EO1 + EO2 treatment; (d) a single nanotube.

Table 2. Roughness parameters (R_a) in the area of 200×200 nm².

Sample	R_a Parameter (nm)
Ti–13Nb–13Zr	25 ± 5
EO1	5 ± 2
EO2	36 ± 11
EO1 + EO2	32 ± 10

The chemical composition of the layers determined by EDS measurements is demonstrated in Table 3. However, because of the oxide volume and thickness examined by the EDS, which exceeds that of the oxide coating of the EO1 sample, the data for this specific case could be result from both thin oxide coating and the alloy. The oxygen content in each coating was determined from stoichiometry, assuming that it formed the stoichiometric oxides.

Table 3. EDS examinations of tested specimens.

Element	EO1 Treatment * wt.%	EO2 Treatment wt.%	EO1 + EO2 Treatment wt.%
O	41.34	56.20	52.61
Ti	45.30	25.98	27.06
F	–	3.50	3.53
Nb	3.21	6.72	9.51
Zr	5.50	5.84	9.02
P	1.12	1.76	1.80

(*) quantities, in this case, must be regarded as only informative.

The Raman spectra of the titanium alloy are shown in Figure 4. According to previous research [66], the Raman spectra for the EO1 coating should display clear signs of the anatase phase four-peak pattern with peaks at 575 cm^{-1} deriving from ν_1 vibrations being the strongest and the other peaks at 144, 198 and 406 cm^{-1} being much weaker. These peaks come from anatase [67,68]. However, here, the small rutile band was observed at 238 and 612 cm^{-1}. For the EO1 + EO2 coating, the intensity of the band 313 cm^{-1} increased and moved to higher frequencies; these peaks originated from the TiO$_2$ band [68]. A very similar situation was noticed for the 198 cm^{-1} peak, which also came from anatase. These findings are in agreement with XRD characterization showing TiO$_2$-specific peaks (for anatase and rutile phases). The Raman spectra of here examined samples are similar to those reported for titanate crystal formed of nanotubes [69].

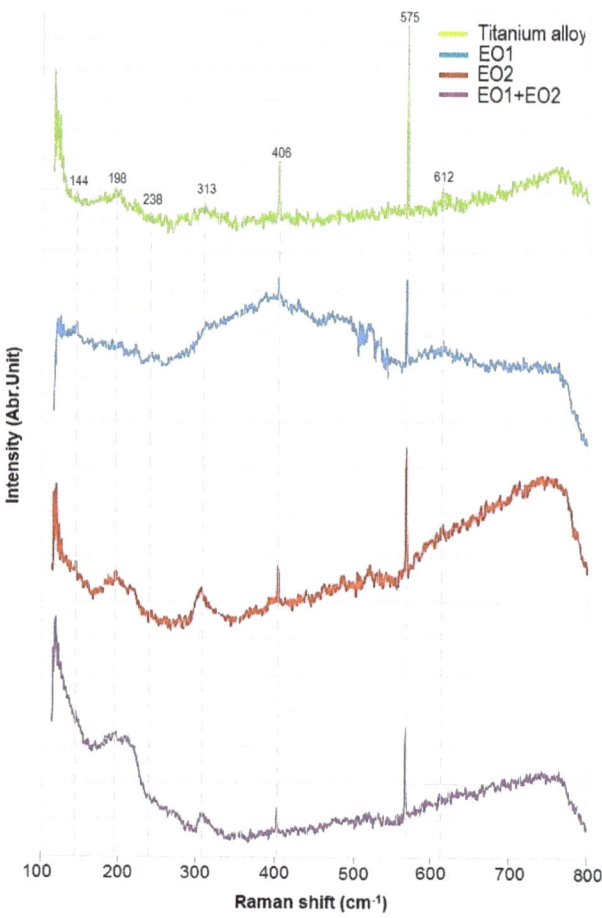

Figure 4. Raman spectra of the oxide coatings for a non-oxidized substrate and after different electrochemical oxidation.

As a result of the GDOES measurements, the values of wavelengths emitted by the excitation of atoms appear for all here present elements, as shown in Figure 5. For the Ti–13Nb–13Zr alloy (Figure 5a), the distribution of elements with erosion (sputtering) time was abrupt and remained at a certain level. For the oxidized EO1 sample (Figure 5b), the maximum intensity for oxygen occurred in the initial phase of the measurement, significantly exceeding the value of the peak derived from titanium, which decayed very quickly. The distribution of the intensity values of particular elements for the EO2 (Figure 5c) and EO1 + EO2 (Figure 5d) coatings was different. In the case of the EO1 + EO2 coating, in the initial phase of the study, there were distinct fluctuations in the intensity of the main alloying elements: Ti, Nb and Zr. The differences in the intensity of the elemental distribution with erosion time are visible, which may be due to different thicknesses of the tested coatings.

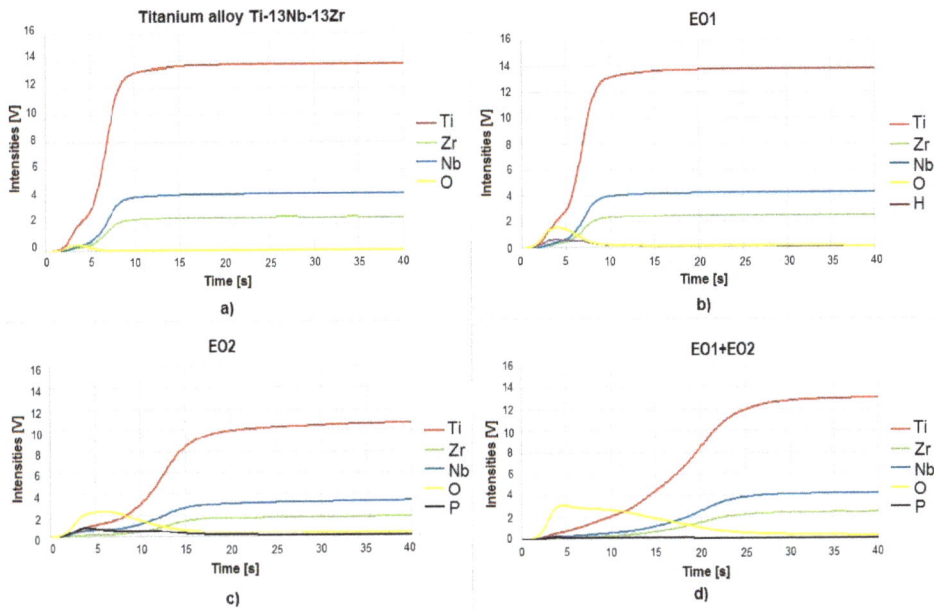

Figure 5. Glow discharge optical emission spectroscopy (GDOES) analysis results. (**a**) Ti–13Nb–13Zr alloy; (**b**) alloy after EO1 treatment; (**c**) alloy after EO2 treatment; (**d**) alloy after EO1 + EO2 treatment.

Analyzing the XRD diagrams (Figure 6) and based on the literature data [70–72], for each sample the characteristic reflexes corresponding to the positions of the α-Ti phase and β-Ti can be found (Figure 6a). Depending on the sample, they differ in intensity and width. In all tested samples, several reflexes from both crystallographic structures were found at appropriate angles.

For oxidized specimens, the reflexes from anatase were observed at three positions, and from rutile–only at one. Due to the overlapping of peaks from titanium and titanium oxide, the crystalline structure of the titania nanotubes could not be determined. A slight decrease in the intensity of the peaks for the EO1 (Figure 6b) sample was observed. In the case of EO1 + EO2 coating (Figure 6d), it became necessary to reduce the reflex intensities of both phases (α-Ti and β-Ti) in comparison with the EO2 coating (Figure 6c). A much more significant decrease in the intensity of the reflexes from the β-Ti phase was observed compared to the reduction of the α-Ti phase, perhaps due to the presence of oxygen. Most likely, the reflection intensity depends on the thickness of the layer; the thicker the layer, the higher the reflex intensity. This result appeared for all oxidized specimens for which a significant increase in reflex intensity was observed. It is worth noting that the primary reflexes from the oxide phase were distinctly the highest for the EO1 + EO2 sample.

3.2. Nanomechanical Properties

The nanohardness, microhardness and Young's modulus values of the specimens are shown in Table 4. The tests showed an increase in both hardness values and Young's modulus for oxidized samples. When considering Young's modulus, the highest value was obtained for the EO1 coating. Similar results were obtained for the nanoparticle layer and hybrid coating.

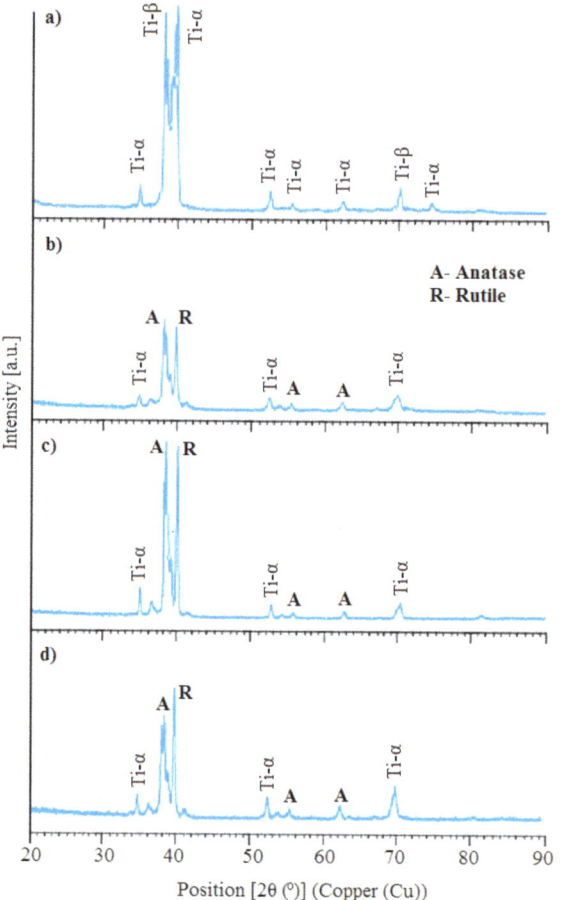

Figure 6. XRD spectra. (**a**) Alloy Ti–13Nb–13Zr; (**b**) alloy after EO1 treatment; (**c**) alloy after EO2 treatment; (**d**) alloy after EO1 + EO2.

Table 4. Mechanical properties of the tested specimens.

Sample	Max. Depth (nm)	Plastic Depth (nm)	Hardness (GPa)	Young's Modulus (GPa)
Ti–13Nb–13Zr	778 ± 17	455 ± 1	0.09 ± 0.06	44.30 ± 0.25
EO1	158 ± 33	152 ± 1	1.01 ± 0.22	65.24 ± 4.13
EO2	138 ± 48	128 ± 1	2.03 ± 0.33	57.30 ± 3.87
EO1 + EO2	241 ± 15	228 ± 4	2.71 ± 0.42	59.73 ± 4.09

An increase in hardness of one order of magnitude was observed for samples with the compact oxide layer. For samples with the EO2 and EO1 + EO2 coatings, the values were similar, much higher compared to polished alloy or the compact layer. The main factor influencing the change in hardness is the thickness of the oxide layer [73–75]. The small thickness of the compact solid layer (80 nm) could cause some errors in the nanoindentation test. It is well known that the response is not only be given by the first indented layer, but the substrate or the subsequent layers may also contribute to the indentation response [76]. In both cases, the thickness of the nanotubular layer (EO2) was so large that there was no response from the substrate.

3.3. Wettability

The results of measurements of water contact angle are presented in Table 5. The decrease of the contact angle was small for the EO1 and significant for the other coatings. The created surfaces were hydrophilic. The most desirable value of contact angle for regeneration applications in hard tissues ranges from 35° to 80° [77].

Table 5. Contact angle for the water droplet for the tested specimens.

Sample	Average Angle (°)
Ti–13Nb–13Zr	83.0 ± 2.4
EO1	78.9 ± 1.7
EO2	29.2 ± 1.4
EO1 + EO2	48.8 ± 1.6

3.4. Corrosion Properties

The corrosion test results are presented in Figure 7. The polarization curves were S-shaped. The decrease in pH value always resulted in a shift of corrosion potential to a more active area. However, the appearance of thin, compact oxide coating slightly worsened the corrosion behavior, and the EO1 + EO2 treatment caused an opposite effect—a shift of the corrosion potential to the more noble area. The corrosion current values can be determined if the Tafel straight lines in potentiodynamic curves (at logarithmic scale) are sufficiently long (at least two decades). Here, this condition has not been fulfilled. Therefore, there is no sufficient base to calculate the corrosion current densities.

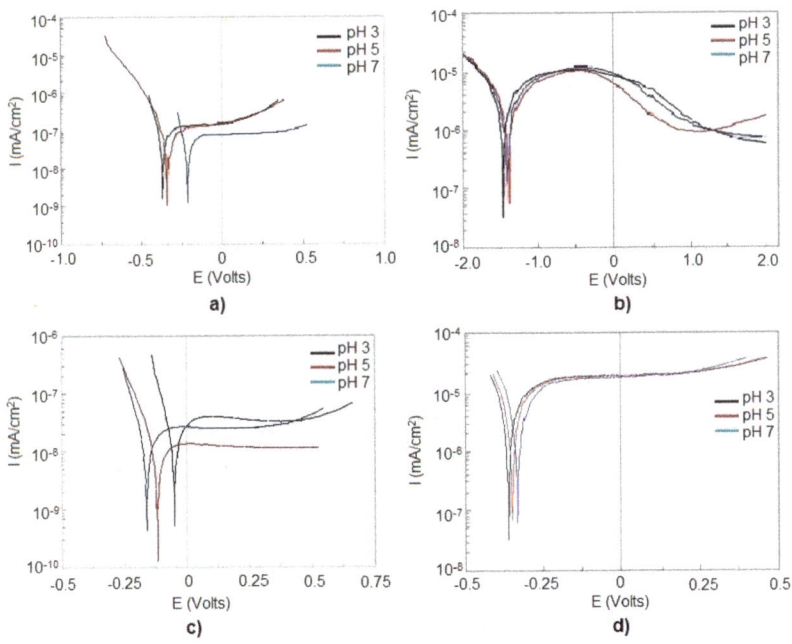

Figure 7. Potentiodynamic polarization curves at different pH. (**a**) Ti–13Nb–13Zr; (**b**) EO1 sample; (**c**) EO2 sample; (**d**) EO1 + EO2 sample.

3.5. Antibacterial Properties

Figure 8 presents the images illustrating the intensity of biofilm formation on the surfaces of materials, measured by absorbance values of solutions of formazan (diluted in DMSO) produced by live cells of bacteria from MTT. After one-day exposure, the lowest absorbance values were observed for the reference and the EO1 samples. The presence of nanotubular surface distinctly increased biofilm formation. In contrast, even the biofilm increased in five days, the lowest levels were attained for the EO2 and EO1 + EO2 samples, for which only a slight difference was noticed between the first and fifth days.

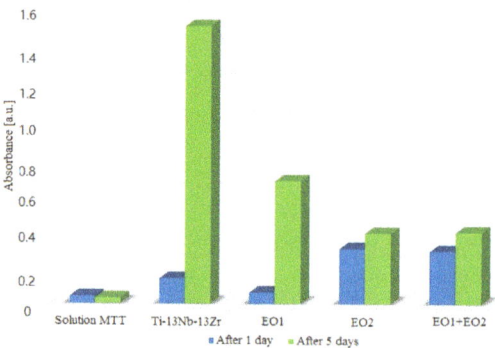

Figure 8. Growth of bacterial film of *Staphylococcus aureus* strain on the tested samples surfaces.

3.6. Cytotoxicity

In these tests (Figure 9), five-day-old cultures formed a reasonably even layer of cells (monolayers), with a large number of figures of mitotic divisions: prophase and metaphase, a small number of polymorphs with no signs of cell damage. The images of cell culture on the surface of the non-oxidized alloy, as well as of the oxidized samples, show that none of the studied surfaces deteriorated the behavior of osteoblasts.

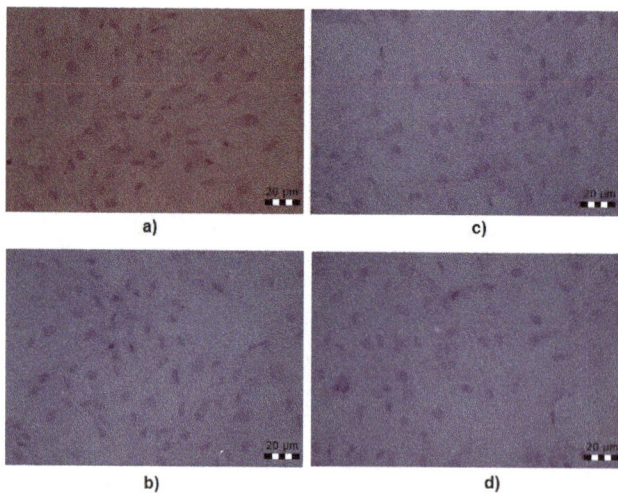

Figure 9. Cytotoxic tests of fibroblastic cells. (**a**) Ti–13Nb–13Zr alloy; (**b**) EO1 sample; (**c**) EO2 sample; (**d**) EO1 + EO2 sample.

4. Discussion

The two-stage oxidation could result in the bi-layer ("sandwich" layer) oxide coating as shown in [47] in which the alloy was subjected to gaseous oxidation and then to the electrochemical oxidation. The last method performed in the presence of HF acid caused the transformation of the upper part of the compact oxide coating into a nanotubular layer, resulting in both highly corrosion-resistant and bioactive coating. However, such a mechanism is possible if oxidation is performed at a high temperature at which oxygen diffusion is fast. On the other hand, with increasing compact oxide thickness, the thickness of the nanotubular layer decreases to zero because of increasing electrical resistance.

Therefore, we have attempted to create the compact oxide layer by the electrochemical method and transform it into a nanotubular layer. The thickness of the oxide coating appearing at room temperature is low because of slow oxygen diffusion, and the thickness of the nanotubular layer creating in the second stage overpasses that of the compact oxide. Therefore, at applied oxidation parameters, the creation of the bilayer coating is not possible. The inner layer shown in Figure 2f may be simply a part of the nanotubular layer, likely, as sometimes observed [71], the interface between the oxide layer and bare metal.

The mechanism of creation of oxide coating during two-stage anodization involves the appearance of a nanotubular layer in the second stage; initially, within the compact oxide layer and then, when the compact layer is fully transformed into the nanotubular structure, within the bare alloy structure. No double structure was observed in any image; however, it affected mechanical and chemical, but not biologic properties.

In the beginning, let us consider the similarities and differences between behavior of specimens subjected to EO2 treatment (nanotubular oxide on bare metal) and EO1 + EO2 oxidation. Neglecting the doubtful inner layer in the EO1 + EO2 sample, it may be said that the form of applied procedure does not affect the coating thickness and roughness. Moreover, the Raman spectra for EO1 + EO2 and EO2 also look very similar in the shape presenting the peaks at the same positions. Raman spectra confirm that the heat treatment of the nanotubes transforms them from the amorphous to the crystalline structure. However, the thickness of the compact oxide layer is much lower than that of the nanotubular layer. It is because the single electrochemical oxidation is determined only by oxygen diffusion, and the growth of the nanotubular layer is much faster as determined by the chemical reaction of etching, being relatively quick. The appearance of the nanotubular structure is an obvious explanation of surface roughness distinctly higher than that of the compact oxide.

The question is, what causes the difference between EO2 is and EO1 + EO2 coatings in some their properties? It may be microstructure, but such results are very difficult to observe. We believe that it is an enrichment with oxygen of the upper zone of the oxide coating after EO1 + EO treatment. The EDS examinations do not confirm this assumption, but their precision is low. However, the GDOES experiments—in which the slower erosion rate and higher oxygen intensity in the EO1 + EO2 coatings compared to EO2 coating—may be an evidence of above proposed phenomenon. Physically, oxygen in the previous oxide layer may occupy also interstices making a microstructure more resistant to diffusion of other elements, more mechanically resistant and more resistant to corrosion by creating the oxides in existing imperfections.

The XRD results are similar for both EO2 and EO1 + EO2 oxidation procedures. The appearance of both rutile and anatase in the oxide layers has been detected what may be surprising as such transformation may occur at temperatures beginning from 450 °C [72], 500 °C [49] or even 550 °C [71] for titanium. Here the heating temperature was 400 °C, close to the suggested beginning of transformation.

The nanoindentation tests show that the presence of nanotubular structure significantly increases hardness and Young's modulus and decreases plastic work. Such results are supported by earlier research [78]. They may result from the specific microstructure of the nanotubular layer, which is comprised of very hard nanotubular oxides, flexible and readily underwent slight deformation. The difference between those values for the EO2 sample and EO1 + EO2 coating may be attributed to the effect of higher oxygen/oxides content in the last sample surface.

The contact angle measurements show an increase in wettability for both EO2 and EO1 + EO2 coatings. The present values permit to classify both surfaces as potentially biocompatible, which may attract proteins and pre-osteoblasts. There have been some assumptions about the best values contact angle for cell attachment assessed at 55° and for bone regeneration from 35° to 80° [70,77]. Here found contact angle values, about 29° for nanotubular layer and 49° for hybrid coating are favorable for biochemical adsorption processes.

The corrosion resistance of titanium alloys is well-known to increase after oxidation. The behavior of the potential beginning from a fast decrease rate followed by a slow rise is typical for a partial stabilization of current density and formation of the highly protective passive film. The current density stabilization suggests the passive film breakdown, in a way similar to what occurs during pitting nucleation and repassivation. Electrochemical potentiodynamic studies have shown how significantly the corrosion quality is affected by the surface quality and thickness of the obtained layers [79,80]. An increase in corrosion resistance of samples covered with amorphous layers (EO1) is visible. The compact, uniform layer provides better corrosion resistance, comparable with spontaneous oxide layers. In the case of EO2 layers (Figure 7c), no decrease in resistance to corrosion was expected. The nanotubes are hollow, and there are voids between them, which were corrosion tunnels - potential places for corrosion development. Saji et al. [81] have observed an increase in the resistance of the oxide layer with a nanotube structure. However, for the Ti–Nb–Ta–Zr alloy, the same authors observed a decrease in corrosion resistance of the nanotubular layer [82]. The thickness of such a layer may play an important role, as it is a barrier to the progression of corrosion [80]. From performed research it follows that a nanotubular layer does not form a suitable protection against corrosion. However, the EO1 + EO2 coating obtained by the electrochemical method shows nobler corrosion potential and likely better corrosion resistance compared to the nanotubular EO2 layer.

The attachment of bacteria to the solid surfaces of different chemical compositions, including alloy surfaces, is affected by the electrostatic double layer, hydrophobicity, roughness and various other factors [83–85]. Bacteria need to overcome the energy barriers to reach the negative energy regions, thereby facilitating the bacterial attachment [86]. In our experiment, we observed a rather moderate development of the population of bacterial cells attached (a bacterial biofilm) to the surfaces of all materials tested after 24 h of incubation. It confirms that some cells of *S. aureus* were able to adsorb on the surfaces of all materials. The MTT assay revealed a bit higher number of bacteria on surfaces of EO2 and EO1 + EO2 compared to EO1 and Ti–13Nb–13Zr. However, no drastic differences were observed between oxidized and non-oxidized specimens in the level of bacterial content after 1st day of incubation.

In contrast, the bacteria development on the non-oxidized alloy drastically increased during the following five days of incubation up to their highest level. In the case of this material, the value of absorbance in MTT assay, which is a consequence of the number of bacteria and the amount of reduced MTT to formazan, reached a value of about 1.5 compared to 0.6, for EO1 and about 0.4 for EO2 and EO1 + EO2. Moreover, in the case of the samples subject to EO2 or EO1 + EO2 oxidation, the levels of bacterial content (biofilms) after one and five days were comparable, which confirms bacterial growth inhibition on the surfaces of these materials which is their essential advantage. The observed influence of the presence of a nanotubular oxide layer on the bacteria attachment or growth inhibition was already reported [87,88]. It may be attributed to the influence of surface topography on the adhesion of bacteria, and it is an evidence that the presence of oxide nanotubes prevents to some extent, thanks to the specific layer microstructure, the danger of bacteria inflammation. The different strategies to avoid infection onto titanium surfaces have been reported: surface modification and coatings by antibiotics, antimicrobial peptides, inorganic antibacterial metal elements and antibacterial polymers [89], but a presence of nanotubular crystalline titanium dioxide also could be useful.

The introduction of nanomaterials and nanostructures may affect the osseointegration processes [90], but also may develop cytotoxicity as several nanomaterials. In our tests (Figure 9), after five days, no cytotoxic effects against the osteoblasts were noticed. These results are following some previous reports [34].

Summarizing it can be said that the application of oxidation joined with etching by HF acid has a different effect on mechanical and chemical properties depending on whether the alloy or oxide coating form the surface zone. The difference can be attributed to the different microstructure within the surface zone, even if the processes leading to such a result cannot be precisely described yet. It may be a saturation of oxides with oxygen or an appearance of more close-packed nanotubes or other not recognized phenomena. This problem will be investigated in detail in the near future.

5. Conclusions

Two-stage Ti–13Nb–13Zr electrochemical oxidation with the use of orthophosphoric acid and subsequently hydrofluoric acid results in an improvement of several nanomechanical and chemical properties such as hardness, Young's modulus and corrosion resistance. No significant effects on biologic properties are observed.

The observed influences may be attributed mainly to the change in the chemical composition and microstructure of the upper zone of the nanotubular layer, inside which the formation of nanotubes occurs not inside the bare alloy, but in the previous anatase layer.

Following the positive effects of present tests, future research will be aimed at recognizing and modeling processes that occur during the formation of titanium oxide nanotubes on previously oxidized alloy, developing the nanotubes of thickness comparable to that of the compact oxide layer, without loss of bioactivity.

Author Contributions: Conceptualization, A.O.; methodology, A.O., J.-M.O., A.W., P.S.; validation, A.O., J.-M.O., P.S.; formal analysis, A.Z.; investigation, A.O., J.-M.O., P.S.; resources, A.O., J.-M.O., A.W., P.S.; writing—original draft preparation, A.O.; writing—review and editing, A.O. and A.Z.; All authors have read and agreed to the published version of the manuscript.

Funding: This research received no external funding.

Acknowledgments: We are grateful to Grzegorz Gajowiec (GUT) for his examinations of oxidized surfaces with the SEM and EDS, Maria Gazda (GUT) for the research and with XRD, Dieter Scharnweber and his Group from Max-Bergmann-Centrum of Biomaterials, Dresden Technical University, for Raman spectroscopy and the research staff of the Institut de Mécanique et d'Ingénierie de Bordeaux for the AFM and GDOES tests.

Conflicts of Interest: The authors declare no conflict of interest.

References

1. Ibrahim, M.Z.; Sarhan, A.A.D.; Yusuf, F.; Hamdi, M. Biomedical materials and techniques to improve the tribological, mechanical and biomedical properties of orthopedic implants—A review article. *J. Alloys Compd.* **2017**, *714*, 636–667. [CrossRef]
2. Oldani, C.; Dominguez, A. Titanium as a biomaterial for implants. *Recent Adv. Arthroplast.* **2012**, 149–162. [CrossRef]
3. Kaur, M.; Singh, K. Review on titanium and titanium based alloys as biomaterials for orthopedic applications. *Mater. Sci. Eng. C* **2019**, *102*, 844–862. [CrossRef] [PubMed]
4. Cordeiro, J.M.; Beline, T.; Ribeiro, A.L.R.; Rangel, E.C.; da Cruz, N.C.; Landers, R.; Faverani, L.P.; Vaz, L.G.; Fais, L.M.G.; Vicente, F.B.; et al. Development of binary and ternary titanium alloys for dental implants. *Dent. Mater.* **2017**, *33*, 1244–1257. [CrossRef]
5. Prasad, S.; Ehrensberger, M.; Prasad Gibson, M.; Kim, H.; Monaco, E.A., Jr. Biomaterial properties of titanium in dentistry. *J. Oral Biosci.* **2015**, *57*, 192–199. [CrossRef]
6. Revathi, A.; Borrás, A.D.; Muñoz, A.I.; Richard, C.; Manivasagam, G. Degradation mechanisms and future challenges of titanium and its alloys for dental implant applications in oral environment. *Mater. Sci. Eng. C* **2017**, *76*, 1354–1368. [CrossRef]
7. Koizumi, H.; Takeuchi, Y.; Imaie, H.; Kawai, T.; Yoneyama, T. Application of titanium and titanium alloys to fixed dental prostheses. *J. Prosthodont. Res.* **2019**, *63*, 266–270. [CrossRef]

8. Acciari, H.A.; Palma, D.P.S.; Codaro, E.N.; Zhou, Q.; Wang, J.; Ling, Y.; Zhang, J.; Zhang, Z. Surface modifications by both anodic oxidation and ion beam implantation on electropolished titanium substrates. *Appl. Surf. Sci.* **2019**, *487*, 1111–1120. [CrossRef]
9. Akatsu, T.; Yamada, Y.; Hoshikawa, Y.; Onoki, T.; Shinoda, Y.; Wakai, F. Multifunctional porous titanium oxide coating with apatite forming ability and photocatalytic activity on a titanium substrate formed by plasma. *Mater. Sci. Eng. C* **2013**, *33*, 4871–4875. [CrossRef]
10. Chen, X.; Chen, Y.; Shen, J.; Xu, J.; Zhu, L.; Gu, X.; He, F.; Wang, H. Positive modulation of osteogenesis on a titanium oxide surface incorporating strontium oxide: An in vitro and in vivo study. *Mater. Sci. Eng. C* **2019**, *99*, 710–718. [CrossRef]
11. Karaji, Z.G.; Hedayati, R.; Pouran, B.; Apachitei, I.; Zadpoor, A.A. Effects of plasma electrolytic oxidation process on the mechanical properties of additively manufactured porous biomaterials. *Mater. Sci. Eng. C* **2017**, *76*, 406–416. [CrossRef] [PubMed]
12. Moravec, H.; Vandrovcova, M.; Chotova, K.; Fojt, J.; Pruchova, E.; Joska, L.; Bacakova, L. Cell interaction with modified nanotubes formed on titanium alloy Ti-6Al-4V. *Mater. Sci. Eng. C* **2016**, *65*, 313–322. [CrossRef] [PubMed]
13. Tanase, C.E.; Golozar, M.; Best, S.M.; Brooks, R.A. Cell response to plasma electrolytic oxidation surface-modified low-modulus β-type titanium alloys. *Colloids Surf. B Biointerfaces* **2019**, *176*, 176–184. [CrossRef]
14. Li, H.; Zhou, D.; Zhang, Q.; Feng, C.; Zheng, W.; He, K.; Lan, Y. Vanadium exposure-induced neurobehavioral alterations among Chinese workers. *Neurotoxicology* **2013**, *36*, 49–54. [CrossRef] [PubMed]
15. Show, C.A.; Tomljenovic, L. Aluminum in the central nervous system (CNS): Toxicity in humans and animals, vaccine adjuvants, and autoimmunity. *Immunol. Res.* **2013**, *56*, 304–316. [CrossRef]
16. Asri, R.I.M.; Harun, W.S.W.; Samykano, M.; Lah, N.A.C.; Ghani, S.A.C.; Tarlochan, F.; Raza, M.R. Corrosion and surface modification on biocompatible metals: A review. *Mater. Sci. Eng. C* **2017**, *77*, 1261–1274. [CrossRef]
17. Park, J.; Bauer, S.; Von Der Mark, K.; Schmuki, P. Nanosize and Vitality: TiO_2 Nanotube Diameter Directs Cell Fate. *Nano Lett.* **2007**, *7*, 1686–1691. [CrossRef]
18. Chen, X.; Fan, H.; Deng, X.; Wu, L.; Yi, T.; Gu, L.; Zhou, C.; Fan, Y.; Zhang, X. Scaffold structural microenvironmental cues to guide tissue regeneration in bone tissue applications. *Nanomaterials* **2018**, *8*, 960. [CrossRef]
19. Dumas, V.; Guignandon, A.; Vico, L.; Mauclair, C.; Zapata, X.; Linossier, M.T.; Bouleftour, W.; Granier, J.; Peyroche, S.; Dumas, J.C.; et al. Femtosecond laser nano/micro patterning of titanium influences mesenchymal stem cell adhesion and commitment. *Biomed. Mater.* **2015**, *10*. [CrossRef]
20. Majkowska, B.; Jazdzewska, M.; Wołowiec, E.; Piekoszewski, W.; Klimek, L.; Zielinski, A. The possibility of use of laser-modified Ti6Al4V alloy in friction pairs in endoprostheses. *Arch. Met. Mater.* **2015**, *60*, 755–758. [CrossRef]
21. Maurel, P.; Weiss, L.; Bocher, P.; Fleury, E.; Grosdidier, T. Oxide dependent wear mechanisms of titanium against a steel counterface: Influence of SMAT nanostructured surface. *Wear* **2019**, *430–431*, 245–255. [CrossRef]
22. Huang, R.; Zhang, L.; Huang, L.; Zhu, J. Enhanced in-vitro osteoblastic functions on β-type titanium alloy using surface mechanical attrition treatment. *Mater. Sci. Eng. C* **2019**, *97*, 688–697. [CrossRef] [PubMed]
23. Ghensi, P.; Bressan, E.; Gardin, C.; Ferroni, L.; Ruffato, L.; Caberlotto, M.; Soldini, C.; Zavan, B. Osteogrowth induction titanium surface treatment reduces ROS production of mesenchymal stem cells increasing their osteogenic commitment. *Mater. Sci. Eng. C* **2017**, *74*, 389–398. [CrossRef] [PubMed]
24. Parchanska-Kowalik, M.; Wołowiec-Korecka, E.; Klimek, L. Effect of chemical surface treatment of titanium on its bond with dental ceramics. *J. Prosthet. Dent.* **2018**, *120*, 470–475. [CrossRef]
25. Asri, R.I.M.; Harun, W.S.W.; Hassan, M.A.; Ghani, S.A.C.; Buyong, Z. A review of hydroxyapatite-based coating techniques: Sol–gel and electrochemical depositions on biocompatible metals. *J. Mech. Behav. Biomed. Mater.* **2016**, *57*, 95–108. [CrossRef]

26. Harun, W.S.W.; Asri, R.I.M.; Alias, J.; Zulkifli, F.H.; Kadirgama, K.; Ghani, S.A.C.; Shariffuddin, J.H.M. A comprehensive review of hydroxyapatite-based coatings adhesion on metallic biomaterials. *Ceram. Int.* **2018**, *44*, 1250–1268. [CrossRef]
27. Bral, A.; Mommaerts, M.Y. In vivo biofunctionalization of titanium patient-specific implants with nano hydroxyapatite and other nano calcium phosphate coatings: A systematic review. *J. Cranio-Maxillofac. Surg.* **2016**, *44*, 400–412. [CrossRef]
28. Supernak-Marczewska, M.; Ossowska, A.; Strąkowska, P.; Zieliński, A. Nanotubular oxide layers and hydroxyapatite coatings on porous titanium alloy Ti13Nb13Zr. *Adv. Mater. Sci.* **2018**, *18*, 17–23. [CrossRef]
29. Bartmanski, M.; Cieslik, B.; Glodowska, J.; Kalka, P.; Pawlowski, L.; Pieper, M.; Zielinski, A. Electrophoretic deposition (EPD) of nanohydroxyapatite—Nanosilver coatings on Ti13Zr13Nb alloy. *Ceram. Int.* **2017**, *43*, 11820–11829. [CrossRef]
30. Li, D.; Li, K.; Shan, H. Improving biocompatibility of titanium alloy scaffolds by calcium incorporated silicalite-1 coatings. *Inorg. Chem. Commun.* **2019**, *102*, 61–65. [CrossRef]
31. Karimi, N.; Kharaziha, M.; Raeissi, K. Electrophoretic deposition of chitosan reinforced graphene oxide-hydroxyapatite on the anodized titanium to improve biological and electrochemical characteristics. *Mater. Sci. Eng. C* **2019**, *98*, 140–152. [CrossRef] [PubMed]
32. Fathyunes, L.; Khalil-Allafi, J.; Moosavifar, M. Development of graphene oxide/calcium phosphate coating by pulse electrodeposition on anodized titanium: Biocorrosion and mechanical behavior. *J. Mech. Behav. Biomed. Mater.* **2019**, *90*, 575–586. [CrossRef]
33. Szklarska, M.; Dercz, G.; Simka, W.; Łosiewicz, B. Ac impedance study on the interfacial properties of passivated Ti13Zr13Nb alloy in physiological saline solution. *Surf. Interface Anal.* **2014**, *46*, 698–701. [CrossRef]
34. Pradhan, D.; Wren, A.W.; Misture, S.T.; Mellott, N.P. Investigating the structure and biocompatibility of niobium and titanium oxides as coatings for orthopedic metallic implants. *Mater. Sci. Eng. C* **2016**, *58*, 918–926. [CrossRef] [PubMed]
35. Gao, A.; Hang, R.; Bai, L.; Tang, B.; Chu, P.K. Electrochemical surface engineering of titanium-based alloys for biomedical application. *Electrochim. Acta* **2018**, *271*, 699–718. [CrossRef]
36. Aniołek, K.; Kupka, M.; Barylski, A. Sliding wear resistance of oxide layers formed on a titanium surface during thermal oxidation. *Wear* **2016**, *356–357*, 23–29. [CrossRef]
37. Khodaei, M.; Kelishadi, S.H. The effect of different oxidizing ions on hydrogen peroxide treatment of titanium dental implant. *Surf. Coat. Technol.* **2018**, *353*, 158–162. [CrossRef]
38. Łęcka, K.M.; Gąsiorek, J.; Mazur-Nowacka, A.; Szczygieł, B.; Antończak, A.J. Adhesion and corrosion resistance of laser-oxidized titanium in potential biomedical application. *Surf. Coat. Technol.* **2019**, *366*, 179–189. [CrossRef]
39. Lin, D.J.; Fuh, L.J.; Chen, C.Y.; Chen, W.C.; Lin, J.H.C.; Chen, C.C. Rapid nano-scale surface modification on micro-arc oxidation coated titanium by microwave-assisted hydrothermal process. *Mater. Sci. Eng. C* **2019**, *95*, 236–247. [CrossRef]
40. He, X.; Zhang, X.; Wang, X.; Qin, L. Review of Antibacterial Activity of Titanium-Based Implants' Surfaces Fabricated by Micro-Arc Oxidation. *Coatings* **2017**, *7*, 45. [CrossRef]
41. Lim, S.-G.; Choe, H.-C. Bioactive apatite formation on PEO-treated Ti-6Al-4V alloy after 3rd anodic titanium oxidation. *Appl. Surf. Sci.* **2019**, *484*, 365–373. [CrossRef]
42. Cordeiro, J.M.; Nagay, B.E.; Ribeiro, A.L.R.; da Cruz, N.C.; Rangel, E.C.; Fais, L.M.G.; Vaz, L.G.; Barão, V.A.R. Functionalization of an experimental Ti-Nb-Zr-Ta alloy with a biomimetic coating produced by plasma electrolytic oxidation. *J. Alloys Compd.* **2019**, *770*, 1038–1048. [CrossRef]
43. Li, Y.; Wang, W.; Liu, H.; Lei, J.; Zhang, J.; Zhou, H.; Qi, M. Formation and in vitro/in vivo performance of "cortex-like" micro/nanostructured TiO_2 coatings on titanium by micro-arc oxidation. *Mater. Sci. Eng. C* **2018**, *87*, 90–103. [CrossRef]
44. Wu, B.; Xiong, S.; Guo, Y.; Chen, Y.; Huang, P.; Yang, B. Tooth-colored bioactive titanium alloy prepared with anodic oxidation method for dental implant application. *Mater. Lett.* **2019**, *248*, 134–137. [CrossRef]
45. Ossowska, A.; Sobieszczyk, S.; Supernak, M.; Zielinski, A. Morphology and properties of nanotubular oxide layer on the "Ti–13Zr–13Nb" alloy. *Surf. Coat. Technol.* **2014**, *258*, 1239–1248. [CrossRef]

46. Li, T.; Gulati, K.; Wang, N.; Zhang, Z.; Ivanovski, S. Understanding and augmenting the stability of therapeutic nanotubes on anodized titanium implants. *Mater. Sci. Eng. C* **2018**, *88*, 182–195. [CrossRef]
47. Ossowska, A.; Beutner, R.; Scharnweber, D.; Zieliński, A. Properties of composite oxide layers on the Ti13Nb13Zr alloy. *Surf. Eng.* **2017**, *33*, 841–848. [CrossRef]
48. Wang, G.; Wan, Y.; Ren, B.; Liu, Z. Bioactivity of micropatterned TiO_2 nanotubes fabricated by micro-milling and anodic oxidation. *Mater. Sci. Eng. C* **2019**, *95*, 114–121. [CrossRef]
49. Roy, P.; Berger, S.; Schmuki, P. TiO_2 Nanotubes: Synthesis and applications. *Angew. Chem. Int. Ed.* **2011**, *50*, 2904–2939. [CrossRef]
50. Beranek, R.; Hildebrand, H.; Schmuki, P. Self-organized porous titanium oxide prepared in H_2SO_4/HF electrolytes. *Electrochem. Solid State Lett.* **2003**, *6*, B12–B14. [CrossRef]
51. Valota, A.T.; LeClere, D.J.; Skeldon, P.; Curioni, M.; Hashimoto, T.; Berger, S.; Kunze, J.; Schmuki, P.; Thompson, G.E. Influence of water content on nanotubular anodic titania formed in fluoride/glycerolelectrolytes. *Electrochim. Acta* **2009**, *54*, 4321–4327. [CrossRef]
52. Albu, S.P.; Ghicov, A.; Aldabergenova, S.; Drechsel, P.; Le Clere, D.; Thompson, G.E.; Macak, J.M.; Schmuki, P. Formation of double-walled TiO_2 nanotubes and robust anatase membranes. *Adv. Mater.* **2008**, *20*, 4135–4139.
53. Habazaki, H.; Fushimi, K.; Shimizu, K.; Skeldon, P.; Thompson, G.E. Fast migration of fluoride ions in growing anodic titanium oxide. *Electrochem. Commun.* **2007**, *9*, 1222–1227. [CrossRef]
54. Berger, S.; Kunze, J.; Schmuki, P.; Valota, A.T.; LeClere, D.J.; Skeldon, P.; Thompson, G.E. Influence of water content on the growth of anodic TiO_2 nanotubes in fluoride-containing ethylene glycol electrolytes. *J. Electrochem. Soc.* **2010**, *157*. [CrossRef]
55. Majchrowicz, A.; Roguska, A.; Pisarek, M.; Lewandowska, M. Tailoring the morphology of nanotubular oxide layers on Ti-24Nb-4Zr-8Sn β-phase titanium alloy. *Thin Solid Films* **2019**, *679*, 15–21. [CrossRef]
56. Huang, J.; Zhang, X.; Yan, W.; Chen, Z.; Shuai, X.; Wang, A.; Wang, Y. Nanotubular topography enhances the bioactivity of titanium implants. *Nanomedicine* **2017**, *13*, 1913–1923. [CrossRef]
57. Pruchova, E.; Kosova, M.; Fojt, J.; Jarolimova, P.; Jablonska, E.; Hybasek, V.; Joska, L. A two-phase gradual silver release mechanism from a nanostructured TiAlV surface as a possible antibacterial modification in implants. *Bioelectrochemistry* **2019**, *127*, 26–34. [CrossRef]
58. Oliveira, W.F.; Arruda, I.R.S.; Silva, G.M.M.; Machado, G.; Coelho, L.C.B.B.; Correia, M.T.S. Functionalization of titanium dioxide nanotubes with biomolecules for biomedical applications. *Mater. Sci. Eng. C* **2017**, *81*, 597–606. [CrossRef]
59. Zhou, J.; Frank, M.A.; Yang, Y.; Boccaccini, A.R.; Virtanen, S. A novel local drug delivery system: Superhydrophobic titanium oxide nanotube arrays serve as the drug reservoir and ultrasonication functions as the drug release trigger. *Mater. Sci. Eng. C* **2018**, *82*, 277–283. [CrossRef]
60. Wu, H.; Xie, L.; Zhang, R.; Tian, Y.; Liu, S.; He, M.; Huang, C.; Tian, W. A novel method to fabricate organic-free superhydrophobic surface on titanium substrates by removal of surface hydroxyl groups. *Appl. Surf. Sci.* **2019**, *479*, 1089–1097. [CrossRef]
61. Vilardella, A.M.; Cinca, N.; Garcia-Giralt, N.; Müller, C.; Dosta, S.; Sarret, M.; Cano, I.G.; Nogués, X.; Guilemany, J.M. In-vitro study of hierarchical structures: Anodic oxidation and alkaline treatments onto highly rough titanium cold gas spray coatings for biomedical applications. *Mater. Sci. Eng. C* **2018**, *91*, 589–596. [CrossRef] [PubMed]
62. Esmaeilnejad, A.; Mahmoudi, P.; Zamanian, A.; Mozafari, M. Synthesis of titanium oxide nanotubes and their decoration by MnO nanoparticles for biomedical applications. *Ceram. Int.* **2019**, *45*, 19275–19282. [CrossRef]
63. Veronesi, F.; Giavaresi, G.; Fini, M.; Longo, G.; Longo, G.; Ioannidu, C.A.; Scotto d'Abusco, A.; Superti, F.; Panzini, G.; Misiano, C.; et al. Osseointegration is improved by coating titanium implants with a nanostructured thin film with titanium carbide and titanium oxides clustered around graphitic carbon. *Mater. Sci. Eng. C* **2017**, *70*, 264–271. [CrossRef] [PubMed]
64. Berbel, L.O.; Bonczek, E.P.; Karousis, I.K.; Kotsakis, G.A.; Costa, I. Determinants of corrosion resistance of Ti-6Al-4V alloy dental implants in an In Vitro model of peri-implant inflammation. *PLoS ONE* **2019**. [CrossRef]
65. Van Gilsa, S.; Masta, P.; Stijnsb, E.; Terryna, H. Colour properties of barrier anodic oxide films on aluminium and titanium studied with total reflectance and spectroscopic ellipsometry. *Surf. Coat. Technol.* **2004**, *185*, 303–310. [CrossRef]

66. Yan, X.; Chen, X. Titanium dioxide nanomaterials. In *Encyclopedia of Inorganic and Bioinorganic Chemistry*; John Wiley & Sons, Ltd.: New York, NY, USA, 2015.
67. Ekoi, E.J.; Gowen, A.; Dorrepaal, R.; Dowling, D.P. Characterisation of titanium oxide layers using Raman spectroscopy and optical profilometry: Influence of oxide properties. *Results Phys.* **2019**, *12*, 1574–1585. [CrossRef]
68. Gajović, A.; Friščić, I.; Plodinec, M.; Iveković, D. High temperature Raman spectroscopy of titanate nanotubes. *J. Mol. Struct.* **2009**, *924–926*, 183–191. [CrossRef]
69. Bavykin, D.V.; Walsh, F.C. Titanate and titania nanotubes: Synthesis, properties and applications. *RSC Nanosci. Nanotechnol.* **2010**, *12*, 12. [CrossRef]
70. Han, B.; Nezhad, E.Z.; Musharavati, F.; Jaber, F.; Bae, S. Tribo-Mechanical Properties and Corrosion Behavior Investigation of Anodized Ti–V Alloy. *Coatings* **2018**, *8*, 459. [CrossRef]
71. Kodama, A.; Bauer, S.; Komatsu, A.; Asoh, H.; Ono, S.; Schmuki, P. Bioactivation of titanium surfaces using coatings of TiO_2 nanotubes rapidly pre-loaded with synthetic hydroxyapatite. *Acta Biomater.* **2009**, *5*, 2322–2330. [CrossRef]
72. Mazare, A.; Totea, G.; Burnei, C.; Schmuki, P.; Demetrescu, I.; Ionita, D. Corrosion, antibacterial activity and haemocompatibility of TiO_2 nanotubes as a function of their annealing temperature. *Corros. Sci.* **2016**, *103*, 215–222. [CrossRef]
73. Hryniewicz, T.; Rokosz, K.; Valíček, J.; Rokicki, R. Effect of magnetoelectropolishing on nanohardness and Young's modulus of titanium biomaterial. *Mater. Lett.* **2012**, *83*, 69–72. [CrossRef]
74. Ficher-Cripps, A.C. Critical Review of Analysis and Interpretation of nanoindentation test data. *Surf. Coat. Technol.* **2006**, *200*, 4153–4165. [CrossRef]
75. Tuck, J.R.; Korsunsky, A.M.; Bhat, D.G.; Bull, S.J. Indentation hardness evaluation of cathodic arc deposited thin hard coatings. *Surf. Coat. Technol.* **2001**, *139*, 63–74. [CrossRef]
76. Jiménez-Piquéa, E.; Gaillardb, Y.; Anglada, M. Instrumented indentation of layered ceramic materials. *Key Eng. Mater.* **2007**, *333*, 107–116. [CrossRef]
77. Hirvonen, J.K. *Ion Implantation*; Academic Press: New York, NY, USA, 1980.
78. Heise, S.; Höhlinger, M.; Hernandez, Y.T.; Palacio, J.J.P.; Ortiz, J.A.R.; Wagener, V.; Virtanen, S.; Boccaccini, A.R. Electrophoretic deposition and characterization of chitosan/bioactive glass composite coatings on Mg alloy substrates. *Electrochim. Acta* **2017**, *232*, 456–464. [CrossRef]
79. Ion, R.; Stoian, A.B.; Dumitriu, C.; Grigorescu, S.; Mazare, A.; Cimpean, A.; Demetrescu, I.; Schmuki, P. Nanochannels formed on TiZr alloy improvebiological response. *Acta Biomater.* **2015**, *24*, 370–377. [CrossRef]
80. Ammar, Y.; Swailes, D.C.; Bridgens, B.N.; Chen, J. Influence of surface roughness on the initial formation of biofilm. *Surf. Coat. Technol.* **2015**, *284*, 410–416. [CrossRef]
81. Saji, V.S.; Choe, H.C.; Brantley, W.A. An electrochemical study on self-ordered nanoporous and nanotubular oxide on Ti-35Nb-5Ta-7Zr alloy for biomedical applications. *Acta Biomater.* **2009**, *5*, 2303–2310. [CrossRef]
82. Mazare, A.; Dilea, M.; Ionita, D.; Demetrescu, I. Electrochemical behaviour in simulated body fluid of TiO_2 nanotubes on TiAlNb alloy elaborated in various anodizing electrolyte. *Surf. Interface Anal.* **2014**, *46*, 186–192. [CrossRef]
83. Lorenzetti, M.; Dogsa, I.; Stosicki, T.; Stopar, D.; Kalin, M.; Kobe, S.; Novak, S. The influence of surface modification on bacterial adhesion to titanium-based substrates. *ACS Appl. Mater. Interfaces* **2015**, *7*, 1644–1651. [CrossRef]
84. Yoda, I.; Koseki, H.; Tomita, M.; Shida, T.; Horiuchi, H.; Sakoda, H.; Osaki, M. Effect of surface roughness of biomaterials on Staphylococcus epidermidis adhesion. *BMC Microbiol.* **2014**, *14*, 234. [CrossRef] [PubMed]
85. Cao, Y.; Su, B.; Chinnaraj, S.; Jana, S.; Bowen, L.; Charlton, S.; Duan, P.; Jakubovics, N.S.; Chen, J. Nanostructured titanium surfaces exhibit recalcitrance towards Staphylococcus epidermidis biofilm formation. *Sci. Rep.* **2018**, *8*, 1071. [CrossRef]
86. Ercan, B.; Kummer, K.M.; Tarquinio, K.M.; Webster, T.J. Decreased Staphylococcus aureus biofilm growth on anodized nanotubular titanium and the effect of electrical stimulation. *Acta Biomater.* **2011**, *7*, 3003–3012. [CrossRef] [PubMed]
87. Simi, V.S.; Rajendran, N. Influence of tunable diameter on the electrochemical behavior and antibacterial activity of titania nanotube arrays for biomedical applications. *Mater. Charact.* **2017**, *129*, 67–79. [CrossRef]
88. Chouirfa, H.; Bouloussa, H.; Migonney, V.; Falentin-Daudré, C. Review of titanium surface modification techniques and coatings for antibacterial applications. *Acta Biomater.* **2019**, *83*, 37–54. [CrossRef]

89. Ahmed, W.; Zhai, Z.; Gao, C. Adaptive antibacterial biomaterial surfaces and their applications. *Mater. Today Bio* **2019**, *2*, 100017. [CrossRef]
90. Zhao, L.; Wang, H.; Huo, K.; Zhang, X.; Wang, W.; Zhang, Y.; Wu, Z.; Chu, P.K. The osteogenic activity of strontium loaded titania nanotube arrays on titanium substrates. *Biomaterials* **2013**, *34*, 19–29. [CrossRef]

© 2020 by the authors. Licensee MDPI, Basel, Switzerland. This article is an open access article distributed under the terms and conditions of the Creative Commons Attribution (CC BY) license (http://creativecommons.org/licenses/by/4.0/).

Article

Nanotubular Oxide Layer Formed on Helix Surfaces of Dental Screw Implants

Magdalena Jażdżewska * and Michał Bartmański

Faculty of Mechanical Engineering and Ship Technology, Gdańsk University of Technology, Narutowicza 11/12, 80-233 Gdańsk, Poland; michal.bartmanski@pg.edu.pl
* Correspondence: magdalena.jazdzewska@pg.edu.pl; Tel.: +48-58-347-17-96

Received: 11 December 2020; Accepted: 18 January 2021; Published: 20 January 2021

Abstract: Surface modification is used to extend the life of implants. To increase the corrosion resistance and improve the biocompatibility of metal implant materials, oxidation of the Ti-13Nb-13Zr titanium alloy was used. The samples used for the research had the shape of a helix with a metric thread, with their geometry imitating a dental implant. The oxide layer was produced by a standard electrochemical method in an environment of 1M H_3PO_4 + 0.3% HF for 20 min, at a constant voltage of 30 V. The oxidized samples were analyzed with a scanning electron microscope. Nanotubular oxide layers with internal diameters of 30–80 nm were found. An analysis of the surface topography was performed using an optical microscope, and the Sa parameter was determined for the top of the helix and for the bottom, where a significant difference in value was observed. The presence of the modification layer, visible at the bottom of the helix, was confirmed by analyzing the sample cross-sections using computed tomography. Corrosion tests performed in the artificial saliva solution demonstrated higher corrosion current and less noble corrosion potential due to incomplete surface coverage and pitting. Necessary improved oxidation parameters will be applied in future work.

Keywords: nanotubular oxide; helix surfaces; dental implants; roughness; corrosion properties

1. Introduction

Titanium and its alloys are nowadays among the most popular biomaterials, called the "gold standard" for endosseous dental implants, even if some adverse reactions may be expected. They possess a lot of important properties, such as their low density, suitable fatigue strength, Young's modulus and specific tensile strength, high resistance to brittle cracking, high corrosion resistance, and the best biocompatibility. Despite that, titanium and its alloys need surface modifications for early osseointegration [1,2]. The type of commercial implant determines surface topography and differences in geometry [3].

Surface modification is nowadays an obligatory treatment of dental implants. Bioactivity of the surface resulting in adhesion of osteoblasts and bone ingrowth can be achieved by the development of surface roughness, creation of bioactive films, and deposition of coatings [2]. Many different methods have been used to change the surface roughness of dental implants, including mechanical techniques such as grinding, polishing, machining, sandblasting and attrition, chemical etching in acids, alkali and fluorides, electrophoretic deposition, and laser treatments [4–11].

The interaction of cells and adsorption of proteins depends on surface structure and is significant in the presence of nanometric pores, which increase the rate of osseointegration and biomechanical fixation [2,7,8,12–17]. A significantly higher bone contact of 27% ($p < 0.05$) was observed in nanotextured compared to machined implants [18]. However, reproducibility of nanoscale surface profiles of titanium with chemical modifications such as acid-etching is quite difficult to achieve and unreliable, and knowledge on the ideal surface roughness parameters for rapid osseointegration is still lacking [19,20].

Among various surface treatments, artificial oxidation seems particularly plausible for titanium dental implants resulting in high corrosion resistance and biocompatibility [21]. The oxidation required for dental implants is currently mostly applied by micro-arc oxidation (MAO) [22,23] and electrochemical oxidation [24]. Gaseous oxidation has also been proposed [25]. MAO induced titanium oxide formation in the anatase crystalline phase and also incorporated Ca, P, and Mg in the film [26–33]. An oxide thickness of 600–1000 nm demonstrated significantly stronger bone responses than that of 17 or 200 nm [30]. The coatings comprising nano TiO_2 and nanohydroxyapatite (nanoHAp) demonstrated a torque value of coated screws significantly greater than that of nanoHAp covered screws [34].

The creation of nanotubular oxide layers on titanium and its alloys is well-known. The formation of nanotubular oxide structures on dental implants has not often been investigated and developed. The overly short life of dental implants observed proves the ineffectiveness of the applied surface modifications and provides prompts for further research. The anodization depends to a great extent on the geometry and structure of the surfaces involved. Indeed, the formation of titanium dioxide nanotubes on flat titanium surfaces, provided by well-known suppliers, does not have the same effect on titanium implants, mainly due to the geometry of the implant, which changes the priority, intensity, and interconnection of the electrochemical processes [35]. Nanotubular oxide layers have been reported to increase the bioactivity of titanium implants [36–38], the nucleation and growth of hydroxyapatite coatings [39], and to introduce antibacterial effects after loading the nanotubes with drugs [40,41]. Such a type of surface was already fabricated on the nontoxic Ti-13Zr-13Nb alloy investigated here [42,43]. This research was aimed at an assessment of the creation of nanotubular oxide layers on screw fixed dental implants and the characterization of the layers obtained on the tops and bottoms of helices of implants.

2. Materials and Methods

2.1. Material

The biphase $\alpha + \beta$ Ti-13Zr-13Nb alloy (SeaBird Metal Materials Co., Baoji, China) with chemical composition presented in Table 1 was investigated in the as-received state.

Table 1. The chemical composition of the Ti-13Zr-13Nb alloy, wt.%. (based on the manufacturer's certificate).

Element	Zr	Nb	O	C	N	Ti
wt.%	13.0	13.0	0.11	0.04	0.019	remainder

2.2. Preparation of Specimens

Round specimens of height 9 mm and diameter 8 mm were prepared by precision milling. The metric thread was cut on all specimens. The remaining impurities were cleaned and the surface was prepared by sand blasting with corundum for 15 s. The cleaning was performed at Aesculap Chifa Ltd. in Nowy Tomyśl, Poland. Immediately before oxidation, the specimens were washed in an ultrasonic bath (Sonic 3, Polsonic, Warsaw, Poland) in isopropanol (POCH, 99.8%, Gliwice, Poland) for 10 min, in distilled water for 3 min, and methanol (POCH, 99.8%, Gliwice, Poland) for 10 min.

2.3. Electrochemical Oxidation

The oxidation was performed using a direct current power supply (MCM/SPN110-01C, Shanghai MCP Corp., Shanghai, China). The specimen tested was connected to the power supply as an anode and the Pt electrode was used as a cathode. The electrolytic bath contained a solution of 150 mL of distilled water, 20 mL of 1 M H_3PO_4, and 1.5 mL of 0.3% HF (both from POCH, Gliwice, Poland). The distance between the electrode tested and the Pt electrode was 15 mm. The solution was neither aerated nor deaerated, and non-stirred. The experiments were carried out at ambient temperature.

The experiments were performed at a voltage of 20 v for 30 min based on previously conducted experiments [42].

2.4. SEM Surface Examination

Scanning electron microscopy (SEM JEOL JSM-7800 F, JEOL Ltd., Tokyo, Japan) instrument equipped with EDS chemical analyzer (Edax Inc., Mahwah, NJ, USA).

2.5. Light Microscopy Assessment of Roughness

A light microscope (VHX-7000, Keyence, Osaka, Japan) was applied to examine the surface topography. Roughness parameters determined by the 3D Form Measurement software were applied to the Sa area.

2.6. Computer Tomography

Tomographic images were obtained using Phoenix v/Tome/xs computer tomography (General Electric, Lewistown, PA, USA).

2.7. Corrosion Examinations in Simulated Body Fluid

Corrosion tests were performed by a potentiodynamic method in simulated body fluid (SBF) at a temperature of 38 °C. The electrochemical measurements were achieved by using a potentiostat/galvanostat (Atlas 0531, Atlas Sollich, Gdańsk, Poland). An artificial saliva solution (SBF) was prepared according to EN ISO 10993-15 [44] by dissolving reagent grade chemical $(NH_2)_2CO$ (0.13 gL^{-1}), NaCl (0.7 gL^{-1}), $NaHCO_3$ (1.5 gL^{-1}), Na_2HPO_4 (0.26 gL^{-1}), K_2HPO_4 (0.2 gL^{-1}), KSCN (0.33 gL^{-1}), KCl (1.2 gL^{-1}) (POCH, Gliwice, Poland). The potential was measured vs. a saturated calomel electrode (SCE) located in the Haber-Luggin capillary. As a counter electrode, a standard platinum electrode was used. The test specimen was stabilized in a solution of artificial saliva for 30 min at open circuit potential OCP. The potential change rate was 1 mV/s within a scan range of −2000 to 1000 mV. The solutions were agitated with a magnetic stirrer. Using the Tafel extrapolation method, the corrosion potential (E_{corr}) and corrosion current density (i_{corr}) values were determined.

3. Results and Discussion

3.1. Substrate Specimens

The surface of the non-oxidized alloy is shown in Figure 1 at two different magnifications. The relatively smooth surface and screw lines can be seen.

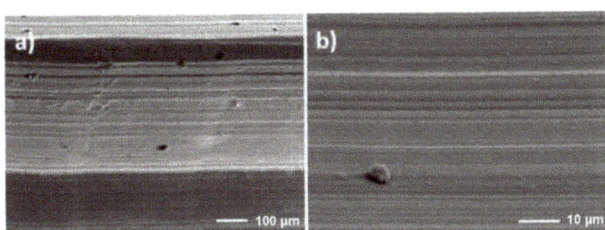

Figure 1. Surface of reference specimen at different magnification: (**a**) 130×, (**b**) 1700×.

3.2. Oxidized Specimens

Figure 2 shows the appearance of the nanotubular oxide layer only in the area at the bottom of the helix. The pores created are spherical and longitudinal. They possess a diameter ranging between 30 and 80 nm. The layer is well adjacent to the substrate and it has a small number of cracks and surface defects. The gradual decrease of the nanoporous layer and its absence at the top of the helix may be attributed to different current densities, different electrochemical potential, and as a consequence a different course of chemical reactions. The current is screened at the bottom at a given potential and the resultant value is sufficient for electrochemical oxidation to occur. At the tops, the current density is too high and the nanotubes formed undergo fast oxidation, its rate exceeding that of chemical dissolution resulting in nanotubes. The current density is higher at the tops of such surfaces, with the effect attributed to the difficult transport of oxygen to and reaction products from this area, and stepwise depolarization of the area close to the bottom followed by a change in open circuit and corrosion potentials. These processes can shift the current and potential values beyond those necessary to form nanotubular oxide layers.

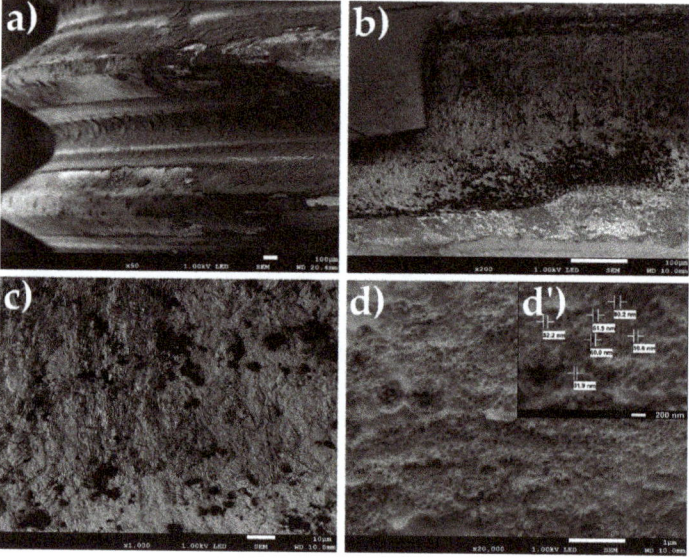

Figure 2. Surfaces of specimens oxidized in an electrochemical way: (**a**) view of specimen, (**b**) surface of helix top, (**c**) surface between top and bottom of the helix, (**d**) bottom of the helix (with different magnifications—d', with the result of measuring the diameter of nanotubes).

The EDS examination results presented in Table 2 suggest the obtaining of a layer of titanium oxide on the surface, which is confirmed mainly by the content of titanium and oxygen. High P content results from the absorption of HPO_4^{2-} anion within the layer pores and it is desired for better bioactivity of the surface. Trace amounts of Ca, K, Fe are observed, which most likely were impurities in the distilled water used.

Results of topography tests are presented in Figure 3. The surface of the top oxide layer is rough and well developed. The roughness profile is 630 nm, the Sa average value is 1.39 ± 0.79 µm on the tip of the helix, and 5.69 ± 2.98 µm on the bottom of the helix (Table 3). Such values in the nanometric range are also useful. The differentiation of the area comprising small nanotubes and rough pores is important. Surface modification led to smoothing the tip of the helix as a result of dissolving roughness peaks. The influence of roughness on oseointegration has been proven. In the case of long-term implants, a positive osteoblast response is required. With increasing roughness, the possibility of osteoblasts settling increases [45]. High roughness also carries the risk of biofilm formation [46]. The topography results confirmed the obtaining of a surface with a high surface roughness value. The lowest values were obtained for the oxidized sample.

Table 2. The EDS examination results of the chemical composition of the oxide layer.

Element	Wt. Pct.	At. Pct.
O	48.49	71.65
Zr	5.72	2.07
P	19.29	14.72
Nb	6.91	1.76
K	0.68	0.41
Ca	0.84	0.49
Ti	17.77	8.77
Fe	0.30	0.13

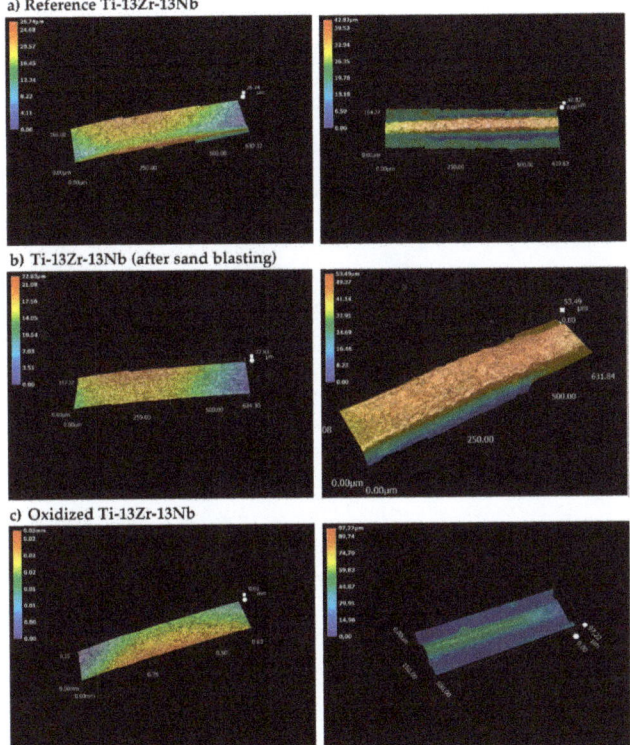

Figure 3. The topography of reference Ti-13Zr-13Nb before (**a**) and after (**b**) sand blasting and oxidized Ti-13Zr-13Nb surfaces (**c**) obtained by light microscopy; the bottom of the helix (**left**) and top (**right**).

Table 3. Sa roughness parameters results.

Specimen	Sa Parameters (µm)	
	Tip of the Helix	Bottom of the Helix
Reference Ti-13Zr13Nb (before sand blasting)	1.81 ± 1.11	9.10 ± 4.62
Ti-13Zr-13Nb (after sand blasting)	1.63 ± 1.40	9.94 ± 5.51
Oxidized Ti-13Zr-13Nb	1.39 ± 0.79	5.69 ± 2.98

The CT investigations showed the appearance of modifications at the bottoms of the helix and not at the tops (Figure 4). The area of modification can be observed as grey and red areas at the bottoms and base alloy as white metal.

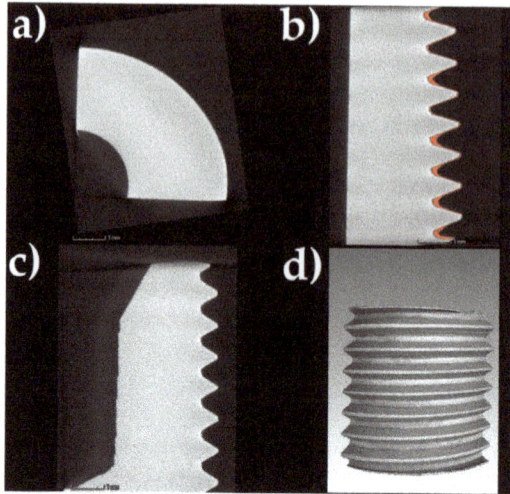

Figure 4. CT images: (**a**) horizontal cross-section with regard to y axis, (**b,c**) vertical cross-sections, (**d**) 3D sample model.

The corrosion results are presented in Figure 5 and Table 4. The creation of the oxide layer became difficult because of much higher current values and a presumed shift of electrochemical potential into more anodic values resulting in the dissolution of metal rather than the oxidation of the surface. The local appearance of the nanotubular and highly rough surface is evidence that some microcells are formed due to change in potential. The local changes in pH value influence the anodization rate, the thickness of the oxide layer and its structure, or even its formation. In case of too low or too high acidity, the oxide layer is unable to achieve the nanotubular structure [47]. Here the anodization was made at the proper HF concentration enhancing the stabilization of the appropriate low pH value and resulting in a short oxidation time, thin nanotubes, a short distance between them, and scarce surface cracks. The roughness of the oxidized surface was close to that observed in similar experiments [48].

The open-circuit potential (OCP) of the non-oxidized specimen was about −199 mV$_{(SCE)}$. The anodic polarization exhibits a narrow plateau between 300 and 1150 mV, which can be attributed to the presence of natural titanium oxide on a specimen surface. The passive current value in this area ranged between 200 and 300 μA. For the previously oxidized specimen, the OCP was about −616 mV. The anodic curve shows a very stable passive region between −200 and 2000 mV. The passive current was about 200 μA in the entire region. However, despite high passive regions, the increase in corrosion current density after oxidation shows that the surface has not been uniformly covered with oxide layers and many microcells could appear in these oxidation conditions. The titanium dioxide layer formed on the surface of the titanium can provide increased corrosion resistance only if it is continuous over the entire surface of the alloy. The layer presented in the paper is characterized by cracks and a lack of continuity. This results in the formation of so-called "corrosion channels", which accelerate the degradation of the material. A similar effect was obtained in research [49]. The occurrence of this phenomenon may explain the deterioration of the corrosive properties compared to the reference sample.

The microscopic investigation reference specimens after corrosion tests showed effects of pitting, some discontinuity of material, and a heterogeneous structure at the bottoms of the helix and at the tops (Figure 6). The microscopic investigation specimens oxidized in an electrochemical way and showed a network of cracks in the surface of the helix top and corrosion pitting in the bottom of the helix (Figure 7), it is probably related to the grater thickness of the obtained modification, which was confirmed by CT tests—Figure 4b.

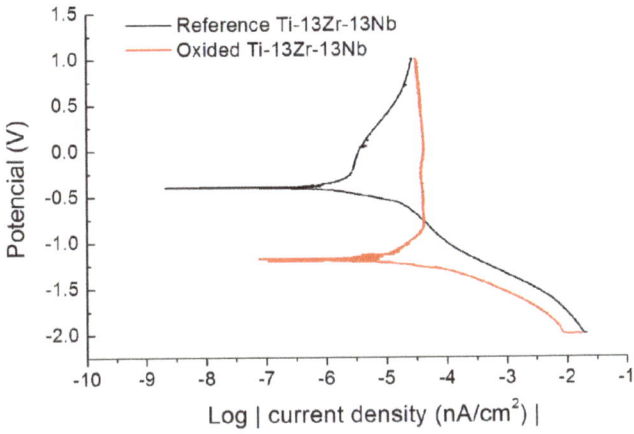

Figure 5. Potentiodynamic polarization curve of reference and oxidized Ti-13Zr-13Nb specimens.

Table 4. Corrosion properties of reference and oxidized Ti-13Zr-13Nb specimens.

Specimen	Current Density (nA/cm^2)	Corrosion Potential (mV)
Reference Ti-13Zr-13Nb	503.25	−392.46
Oxidized Ti-13Zr-13Nb	1451.00	−1174.13

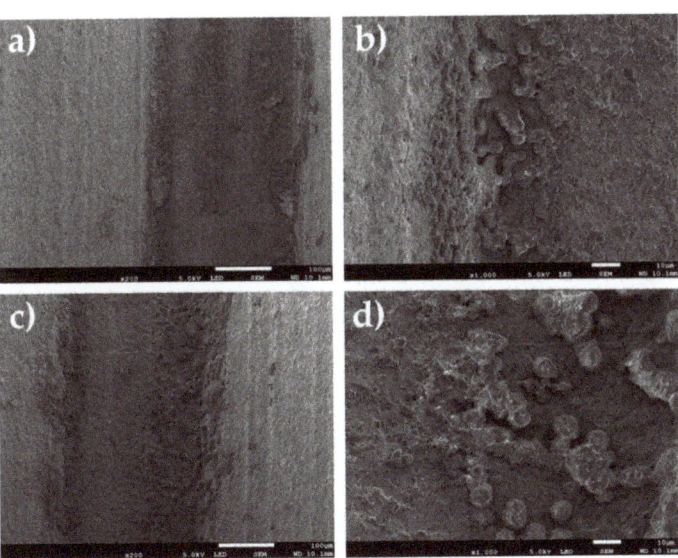

Figure 6. Surface of reference specimen after corrosion test at different magnifications: (**a**) surface of helix top ×200, (**b**) surface of helix top ×1000, (**c**) bottom of the helix ×200, (**d**) bottom of the helix ×1000.

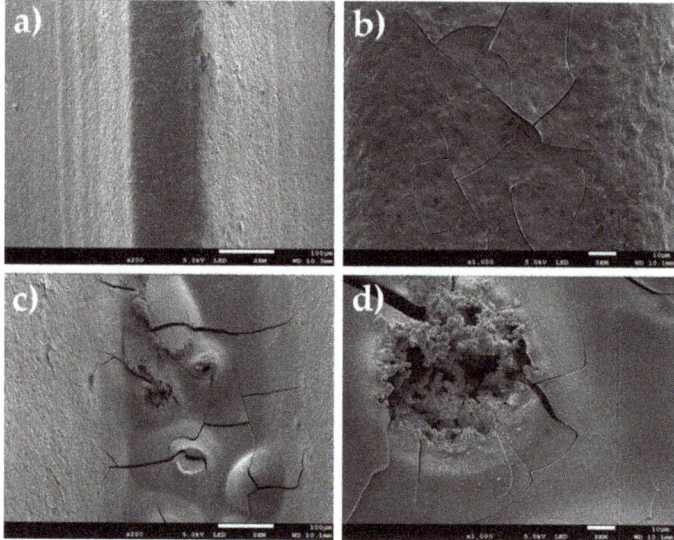

Figure 7. Surfaces of specimens oxidized in an electrochemical way after corrosion tests: (**a**) surface of helix top ×200, (**b**) surface of helix top ×1000, (**c**) bottom of the helix ×200, (**d**) bottom of the helix ×1000.

4. Conclusions

In summary, nanotubular oxidation on the helix lines of titanium dental implants is possible, but it depends heavily on the geometric shape of the implant, anodization parameters, and environment composition. The parameters proposed here make it possible to obtain the nanotubes on the bottom of the helix and distinctly roughen almost all remaining surfaces. However, the applied conditions applied indicate that future investigations be oriented towards oxidation of the whole surface by introducing slightly higher HF contents and lower current values and mixing the electrolyte bath.

Author Contributions: Conceptualization, M.J.; methodology, M.J. and M.B.; formal analysis, M.J. and M.B.; investigation, M.J. and M.B.; writing—original draft preparation, M.J.; writing—review and editing, M.J. and M.B.; supervision, M.J. All authors have read and agreed to the published version of the manuscript.

Funding: This research received no external funding.

Institutional Review Board Statement: Not applicable.

Informed Consent Statement: Not applicable.

Data Availability Statement: The data presented in this study are available on request from the corresponding author.

Acknowledgments: Authors thank the students—M. Get, M. Karczmarczyk, and K. Gąsiorowska for their technical assistance in some tests. The helpful comments of Andrzej Zielinski are gratefully acknowledged.

Conflicts of Interest: The authors declare no conflict of interest.

References

1. Osman, R.B.; Swain, M.V. A critical review of dental implant materials with an emphasis on titanium versus zirconia. *Materials* **2015**, *8*, 932–958. [CrossRef] [PubMed]
2. Le Guéhennec, L.; Soueidan, A.; Layrolle, P.; Amouriq, Y. Surface treatments of titanium dental implants for rapid osseointegration. *Dent. Mater.* **2007**, *23*, 844–854. [CrossRef] [PubMed]

3. Mendoza-Arnau, A.; Vallecillo-Capilla, M.F.; Cabrerizo-Vílchez, M.Á.; Rosales-Leal, J.I. Topographic characterisation of dental implants for commercial use. *Med. Oral Patol. Oral Cir. Bucal* **2016**, *21*, e631–e636. [CrossRef] [PubMed]
4. Kulkarni, M.; Mazare, A.; Schmuki, P.; Iglič, A. Biomaterial Surface Modification Of Titanium and Titanium Alloys for Medical Applications. In *Nanomedicine*; One Central Press: Altrincham, UK, 2014; pp. 111–136.
5. Ellingsen, J.E.; Johansson, C.B.; Wennerberg, A.; Holmen, A. Improved retention and bone-tolmplant contact with fluoride-modified titanium implants. *Int. J. Oral Maxillofac. Implant.* **2004**, *19*, 659–666.
6. Buser, D.; Broggini, N.; Wieland, M.; Schenk, R.K.; Denzer, A.J.; Cochran, D.L.; Hoffmann, B.; Lussi, A.; Steinemann, S.G. Enhanced bone apposition to a chemically modified SLA titanium surface. *J. Dent. Res.* **2004**, *83*, 529–533. [CrossRef] [PubMed]
7. Garg, H.; Bedi, G.; Garg, A. Implant surface modifications: A review. *J. Clin. Diagn. Res.* **2012**, *6*, 319–324.
8. Gehrke, S.A.; Taschieri, S.; Massimo, D.P.; Coelho, P.G. The positive biomechanical effects of titanium oxide for sandblasting implant surface as an alternative to aluminium oxide. *J. Oral Implantol.* **2015**, *41*, 515–522. [CrossRef]
9. Majkowska-Marzec, B.; Tęczar, P.; Bartmański, M.; Bartosewicz, B.; Jankiewicz, B.J. Mechanical and Corrosion Properties of Laser Surface-Treated Ti13Nb13Zr Alloy with MWCNTs Coatings. *Materials* **2020**, *13*, 3991. [CrossRef]
10. Pawłowski, Ł.; Bartmański, M.; Strugała, G.; Mielewczyk-Gryń, A.; Jażdżewska, M.; Zieliński, A. Electrophoretic Deposition and Characterization of Chitosan/Eudragit E 100 Coatings on Titanium Substrate. *Coatings* **2020**, *10*, 607. [CrossRef]
11. Puckett, S.D.; Taylor, E.; Raimondo, T.; Webster, T.J. The relationship between the nanostructure of titanium surfaces and bacterial attachment. *Biomaterials* **2010**, *31*, 706–713. [CrossRef]
12. Scopelliti, P.E.; Borgonovo, A.; Indrieri, M.; Giorgetti, L.; Bongiorno, G.; Carbone, R.; Podestà, A.; Milani, P. The effect of surface nanometre-scale morphology on protein adsorption. *PLoS ONE* **2010**, *5*, e11862. [CrossRef] [PubMed]
13. Zinger, O.; Anselme, K.; Denzer, A.; Habersetzer, P.; Wieland, M.; Jeanfils, J.; Hardouin, P.; Landolt, D. Time-dependent morphology and adhesion of osteoblastic cells on titanium model surfaces featuring scale-resolved topography. *Biomaterials* **2004**, *25*, 2695–2711. [CrossRef] [PubMed]
14. Kohavi, D.; Badihi Hauslich, L.; Rosen, G.; Steinberg, D.; Sela, M.N. Wettability versus electrostatic forces in fibronectin and albumin adsorption to titanium surfaces. *Clin. Oral Implant. Res.* **2013**, *24*, 1002–1008. [CrossRef] [PubMed]
15. Brett, P.M.; Harle, J.; Salih, V.; Mihoc, R.; Olsen, I.; Jones, F.H.; Tonetti, M. Roughness response genes in osteoblasts. *Bone* **2004**, *35*, 124–133. [CrossRef] [PubMed]
16. Webster, T.J.; Ejiofor, J.U. Increased osteoblast adhesion on nanophase metals: Ti, Ti6Al4V, and CoCrMo. *Biomaterials* **2004**, *25*, 4731–4739. [CrossRef] [PubMed]
17. Gittens, R.A.; Olivares-Navarrete, R.; Schwartz, Z.; Boyan, B.D. Implant osseointegration and the role of microroughness and nanostructures: Lessons for spine implants. *Acta Biomater.* **2014**, *10*, 3363–3371. [CrossRef]
18. Pinheiro, F.A.L.; de Almeida Barros Mourão, C.F.; Diniz, V.S.; Silva, P.C.; Meirelles, L.; Junior, E.S.; Schanaider, A. In-vivo bone response to titanium screw implants anodized in sodium sulfate1. *Acta Cir. Bras.* **2014**, *29*, 376–382. [CrossRef]
19. Ogawa, T.; Saruwatari, L.; Takeuchi, K.; Aita, H.; Ohno, N. Ti nano-nodular structuring for bone integration and regeneration. *J. Dent. Res.* **2008**, *87*, 751–756. [CrossRef]
20. Sugita, Y.; Ishizaki, K.; Iwasa, F.; Ueno, T.; Minamikawa, H.; Yamada, M.; Suzuki, T.; Ogawa, T. Effects of pico-to-nanometer-thin TiO2 coating on the biological properties of microroughened titanium. *Biomaterials* **2011**, *32*, 8374–8384. [CrossRef]
21. Liu, X.; Chu, P.K.; Ding, C. Surface modification of titanium, titanium alloys, and related materials for biomedical applications. *Mater. Sci. Eng. R Rep.* **2004**, *47*, 49–121. [CrossRef]
22. Tailor, S.; Rakoch, A.G.; Gladkova, A.A.; Van Truong, P.; Strekalina, D.M.; Sourkouni, G.; Manjunath, S.Y.; Takagi, T. Kinetic features of wear-resistant coating growth by plasma electrolytic oxidation. *Surf. Innov.* **2018**, *6*, 150–158. [CrossRef]

23. Aydogan, D.T.; Muhaffel, F.; Acar, O.K.; Topcuoglu, E.N.; Kulekci, H.G.; Kose, G.T.; Baydogan, M.; Cimenoglu, H. Surface modification of Ti6Al4V by micro-arc oxidation in AgC$_2$H$_3$O$_2$-containing electrolyte. *Surf. Innov.* **2018**, *6*, 277–285. [CrossRef]
24. Webster, T.J.; Yao, C. Anodization: A Promising Nano-Modification Technique of Titanium-based Implants for Orthopedic Appl. In *Surface Engineered Surgical Tools and Medical Devices*; Jackson, M.J., Ahmed, W., Eds.; Springer US: Boston, MA, USA, 2007; pp. 21–47.
25. Li, G.; Qu, S.; Ren, Z.; Li, X. Surface Modification Layer of Ti-6Al-4V Produced By Surface Rolling and Thermal Oxidation. *Surf. Innov.* **2017**, *5*, 1–29. [CrossRef]
26. Sul, Y.T. The significance of the surface properties of oxidized titanium to the bone response: Special emphasis on potential biochemical bonding of oxidized titanium implant. *Biomaterials* **2003**, *24*, 3893–3907. [CrossRef]
27. Sul, Y.T.; Johansson, C.B.; Jeong, Y.; Wennerberg, A.; Albrektsson, T. Resonance frequency and removal torque analysis of implants with turned and anodized surface oxides. *Clin. Oral Implant. Res.* **2002**, *13*, 252–259. [CrossRef] [PubMed]
28. Sul, Y.T.; Johansson, C.B.; Petronis, S.; Krozer, A.; Jeong, Y.; Wennerberg, A.; Albrektsson, T. Characteristics of the surface oxides on turned and electrochemically oxidized pure titanium implants up to dielectric breakdown: The oxide thickness, micropore configurations, surface roughness, crystal structure and chemical composition. *Biomaterials* **2002**, *23*, 491–501. [CrossRef]
29. Sul, Y.-T.; Johansson, C.B.; Albrektsson, T. Oxidized titanium screws coated with calcium ions and their performance in rabbit bone. *Int. J. Oral Maxillofac. Implant.* **2002**, *17*, 625–634.
30. Choi, J.W.; Heo, S.J.; Koak, J.Y.; Kim, S.K.; Lim, Y.J.; Kim, S.H.; Lee, J.B. Biological responses of anodized titanium implants under different current voltages. *J. Oral Rehabil.* **2006**, *33*, 889–897. [CrossRef]
31. Elias, C.N. Titanium dental implant surfaces. *Rev. Mater.* **2010**, *15*, 138–142. [CrossRef]
32. Shayganpour, A.; Rebaudi, A.; Cortella, P.; Diaspro, A.; Salerno, M. Electrochemical coating of dental implants with anodic porous titania for enhanced osteointegration. *Beilstein J. Nanotechnol.* **2015**, *6*, 2183–2192. [CrossRef]
33. Van Vuuren, D.J.; Laubscher, R.F. Surface Friction Behaviour of Anodized Commercially Pure Titanium Screw Assemblies. *Procedia CIRP* **2016**, *45*, 251–254. [CrossRef]
34. Nasir, M.; Abdul Rahman, H. Mechanical Evaluation of Pure Titanium Dental Implants Coated with a Mixture of Nano Titanium Oxide and Nano Hydroxyapatite. *J. Baghdad Coll. Dent.* **2016**, *28*, 38–43. [CrossRef]
35. Portan, D.V.; Nikolopoulou, F.; Bairami, V.; Mouzakis, D.; Papanicolaou, G.C.; Deligianni, D.D. Electrochemical Surface Processing Applied for the Functionalization of Titanium Screw Type Implants. *J. Mater. Sci. Surf. Eng.* **2016**, *4*, 376–382.
36. Huang, J.; Zhang, X.; Yan, W.; Chen, Z.; Shuai, X.; Wang, A.; Wang, Y. Nanotubular topography enhances the bioactivity of titanium implants. *Nanomed. Nanotechnol. Biol. Med.* **2017**, *13*, 1913–1923. [CrossRef]
37. Oliveira, W.F.; Arruda, I.R.S.; Silva, G.M.M.; Machado, G.; Coelho, L.C.B.B.; Correia, M.T.S. Functionalization of titanium dioxide nanotubes with biomolecules for biomedical applications. *Mater. Sci. Eng. C* **2017**, *81*, 597–606. [CrossRef]
38. Weszl, M.; Tóth, K.L.; Kientzl, I.; Nagy, P.; Pammer, D.; Pelyhe, L.; Vrana, N.E.; Scharnweber, D.; Wolf-Brandstetter, C.; Joób, F.Á.; et al. Investigation of the mechanical and chemical characteristics of nanotubular and nano-pitted anodic films on grade 2 titanium dental implant materials. *Mater. Sci. Eng. C* **2017**, *78*, 69–78. [CrossRef]
39. Suchanek, K.; Hajdyła, M.; Maximenko, A.; Zarzycki, A.; Marszałek, M.; Jany, B.R.; Krok, F. The influence of nanoporous anodic titanium oxide substrates on the growth of the crystalline hydroxyapatite coatings. *Mater. Chem. Phys.* **2017**, *186*, 167–178. [CrossRef]
40. Yazici, H.; Habib, G.; Boone, K.; Urgen, M.; Utku, F.S.; Tamerler, C. Self-assembling antimicrobial peptides on nanotubular titanium surfaces coated with calcium phosphate for local therapy. *Mater. Sci. Eng. C* **2019**, *94*, 333–343. [CrossRef]
41. Wang, L.-N.; Jin, M.; Zheng, Y.; Guan, Y.; Lu, X.; Luo, J.-L. Surface modification of metallic implants with anodic oxide nanotubular arrays via electrochemical anodization techniques. In *Nanomedicine*; Seifalian, A., de Mel, A., Kalaskar, D.M., Eds.; One Central Press Ltd.: Manchester, UK, 2015; pp. 313–332.
42. Ossowska, A.; Sobieszczyk, S.; Supernak, M.; Zielinski, A. Morphology and properties of nanotubular oxide layer on the "Ti-13Zr-13Nb" alloy. *Surf. Coat. Technol.* **2014**, *258*, 1239–1248. [CrossRef]

43. Ossowska, A.; Beutner, R.; Scharnweber, D.; Zielinski, A. Properties of composite oxide layers on the Ti13Nb13Zr alloy. *Surf. Eng.* **2017**, *33*, 841–848. [CrossRef]
44. EN ISO 10993-15:2009. *Biological Evaluation of Medical Devices. Identification and Quantification of Degradation Products from Metals and Alloys*; ISO: Geneva, Switzerland, 2009.
45. Ehlert, M.; Radtke, A.; Jędrzejewski, T.; Roszek, K.; Bartmański, M.; Piszczek, P. In Vitro Studies on Nanoporous, Nanotubular and Nanosponge-Like Titania Coatings, with the Use of Adipose-Derived Stem Cells. *Materials* **2020**, *13*, 1574. [CrossRef]
46. Alam, F.; Balani, K. Adhesion force of staphylococcus aureus on various biomaterial surfaces. *J. Mech. Behav. Biomed. Mater.* **2017**, *65*, 872–880. [CrossRef]
47. Macak, J.M.; Schmuki, P. Anodic growth of self-organized anodic TiO_2 nanotubes in viscous electrolytes. *Electrochim. Acta* **2006**, *52*, 1258–1264. [CrossRef]
48. Klimas, J.; Dudek, A.; Klimas, M. Surface Refinement of Titanium Alloy TI6AL4V ELI. *Eng. Biomater.* **2012**, *15*, 52–54.
49. Bartmanski, M.; Zielinski, A.; Majkowska-Marzec, B.; Strugala, G. Effects of solution composition and electrophoretic deposition voltage on various properties of nanohydroxyapatite coatings on the Ti13Zr13Nb alloy. *Ceram. Int.* **2018**, *44*, 19236–19246. [CrossRef]

Publisher's Note: MDPI stays neutral with regard to jurisdictional claims in published maps and institutional affiliations.

© 2021 by the authors. Licensee MDPI, Basel, Switzerland. This article is an open access article distributed under the terms and conditions of the Creative Commons Attribution (CC BY) license (http://creativecommons.org/licenses/by/4.0/).

Article

Effects of Micro-Arc Oxidation Process Parameters on Characteristics of Calcium-Phosphate Containing Oxide Layers on the Selective Laser Melted Ti13Zr13Nb Alloy

Magda Dziaduszewska [1],*, Masaya Shimabukuro [2], Tomasz Seramak [1], Andrzej Zieliński [1] and Takao Hanawa [3]

1. Biomaterials Division, Department of Materials Engineering and Bonding, Gdańsk University of Technology, 80-233 Gdańsk, Poland; tseramak@pg.edu.pl (T.S.); andrzej.zielinski@pg.edu.pl (A.Z.)
2. Department of Biomaterials, Faculty of Dental Science, Kyushu University, 3-1-1 Maidashi, Higashi-ku, Fukuoka 812-8582, Japan; shimabukuro@dent.kyushu-u.ac.jp
3. Institute of Biomaterials and Bioengineering, Tokyo Medical and Dental University, 2-3-10 Kanda-surugadai, Chiyoda-ku, Tokyo 101-0062, Japan; hanawa.met@tmd.ac.jp
* Correspondence: magda.dziaduszewska@pg.edu.pl

Received: 29 June 2020; Accepted: 24 July 2020; Published: 30 July 2020

Abstract: Titania-based films on selective laser melted Ti13Zr13Nb have been formed by micro-arc oxidation (MAO) at different process parameters (voltage, current, processing time) in order to evaluate the impact of MAO process parameters in calcium and phosphate (Ca + P) containing electrolyte on surface characteristic, early-stage bioactivity, nanomechanical properties, and adhesion between the oxide coatings and substrate. The surface topography, surface roughness, pore diameter, elemental composition, crystal structure, surface wettability, and the early stage-bioactivity in Hank's solution were evaluated for all coatings. Hardness, maximum indent depth, Young's modulus, and $E_{coating}/E_{substrate}$, H/E, H^3/E^2 ratios were determined in the case of nanomechanical evaluation while the MAO coating adhesion properties were estimated by the scratch test. The study indicated that the most important parameter of MAO process influencing the coating characteristic is voltage. Due to the good ratio of structural and nanomechanical properties of the coatings, the optimal conditions of MAO process were found at 300 V during 15 min, at 32 mA or 50 mA of current, which resulted in the predictable structure, high Ca/P ratio, high hydrophilicity, the highest demonstrated early-stage bioactivity, better nanomechanical properties, the elastic modulus and hardness well close to the values characteristic for bones, as compared to specimens treated at a lower voltage (200 V) and uncoated substrate, as well as a higher critical load of adhesion and total delamination.

Keywords: titanium alloys; micro-arc oxidation; composite oxide coatings; microstructure; properties

1. Introduction

Titanium and its alloys are widely used for medical applications. Among them, pure Ti (cp-Ti) or Ti with Al, V, and Mo alloying elements are still among the most common materials used for implantation. However, due to the reported mismatch in mechanical properties between bones and those materials, leading to the stress-shielding and implant loosening, as well as the toxicity of Al and V, new solutions are considered. Finally, among the perspective materials are β-type titanium alloys [1–3] known as low modulus and bioinert metals containing Ti, Zr, Nb, Hf, Ta alloying elements [4,5]. Although the material ensures preferable mechanical properties, which has been proved as an effective factor in inhibiting bone atrophy [6] there is still a problem with its bioactivity.

Many articles have reported that the composite layer can increase the surface area and roughness of the implant, and promote the ingrowth of the tissue while improving the adhesion between the bone tissue and implant. There are many methods which can enhance the bone tissue response, such as plasma spray, magnetron sputtering, ion implantation, and anodic treatment [7]. However, micro-arc oxidation (MAO), also called plasma electrochemical oxidation (PEO), has gained special attention due to its applicability to biocompatible coating deposition with a gradient structure, rough and porous morphology on such metals as Ti, Al, Mg, Nb, Zr, and their alloys. According to the literature, the MAO technique combines many advantages and exhibits better improvement of surface properties compared to other conventional methods. Laser surface modification improves the wear and corrosion resistance of Ti alloys, however the coatings exhibit low bond strength and the residual stresses produced during remelting and solidification which lead to many cracks in the laser treated coatings [8]. Similarly, plasma spraying improves biocompatibility, wear resistance, and thermal stability, but the formation of tensile forces between the substrate and coatings cause cracks and weaken the bonds. On the other hand, physical vapour deposition (PVD) and chemical vapour deposition (CVD) brings out the coatings of high density and strong adhesion to the substrate; however, such hard coatings also exhibit a large mismatch between mechanical properties of the substrate and coatings and in consequence may cause delamination. The sol-gel films demonstrate that the low adhesion strength to the substrate, possess a limited bioactivity, and their thickness is difficult to control. The friction stir processing (FSP) method improves biofunctionality, corrosion, and wear resistance, but due to the low processing rate and inferior flexibility, it has limitation for the use on complex-geometry [9]. Conventional electrochemical modification is able to produce a reproducible and well-defined coatings, nevertheless the improvement in hardness and wear resistance of Ti alloys by this technique is seldom reported. Compared to the modification method presented above, the micro-arc oxidation (MAO) technique improves the corrosion and wear resistance of titanium alloys, significantly enhances biocompatibility and bioactivity, provides a better bonding between the substrate and coating, and it has the capability to coat the complex-shape objects what constitutes a big advantage for biomedical applications [10]. The MAO process has been investigated by several research groups in recent years, mainly concerned with the biologic behavior, corrosion behavior [11,12], and the characteristics of the surface layer in terms of crystalline structure, topography, porosity or composition [13]. According to the literature, the coating formed in the electrolyte containing Ca and P ions can significantly improve the bioactivity of titanium implants [14–17]. What is more, the structure and properties of MAO coatings can be easily controlled by different MAO process parameters such as applied voltage, current density, time processing, as well as various electrolyte compositions [18–20]. Du et al. [19] and Wei et al. [13] have reported that the applied voltage has significant influence on the structureal parameters such as thickness, the atomic ratio of Ca to P, micropore number, and size of the MAO film on Ti and Zr1Nb alloy [19] and Ti6Al4V alloy [13]. Komarova et al. [21] has shown that the coating thickness, roughness, and the average size of the structural elements grow linearly with increasing the MAO voltage from 200 to 370 V. While, the MAO time effects only the coating thickness. Another research report [22] described the morphology, elemental composition, phase components, and bioactivity in four types of electrolytic solutions and the applied voltage in the range of 200–500 V. Among the sodium carbonate, sodium phosphate, acetate monohydrate (CA), and a mixture of CA and β-glycerophosphate disodium salt pentahydrate (GP), only the last mixture treated by high voltage could induce apatite on the titanium surface and exhibited bioactivity. Alves et al. [7] also evaluated the MAO process duration and electrolyte composition on topography, morphology, chemical composition, crystalline structure, biological and tribocorrosion behavior of the Ca- and P-coatings on cp-Ti. Finally, Tsutsumi et al. [23,24] have already described the characteristic of MAO coatings in various electrolytes and electrochemical conditions on β-type titanium alloys. The results [23] have shown that the titanium with Nb, Ta, Zr elements (TNTZ) under the MAO treatment (with a positive voltage, constant-current condition of 12 mA, and time processing of 8 min) in a mixture of calcium glycerophosphate and magnesium acetate, results in thick calcium phosphate layers formed on the TNTZ after immersion in Hank's solution. The same research group investigated the structure, hardness, and Young's modulus after the MAO process on Ti15Zr7.5Mo with a positive maximum voltage of 400 V and current density of 31.2 mA/cm^2 applied for 10 min. The electrolyte for MAO treatment was 0.1 mol L^{-1} calcium glycerophosphate and 0.15 mol L^{-1} calcium acetate. The research has shown larger hardness (420 HV) compared to the commonly used Ti-5Al-4V alloy (320 HV) and TNTZ (180 HV), as well as higher Young's modulus (104–112 GPa) compared to TNTZ (80 GPa) [24]. Similarly, Wang et al. [25] while studying the differences of MAO coating on Ti6Al4V and Ti35Nb2Ta3Zr, proved that titanium with Zr and Nb alloying elements possessed a better characteristic in terms of excellent corrosion resistance, hydrophilic, and film-forming properties. It can be noticed, that the most previous research with the effect of MAO process parameters on oxidized coatings has been mostly focused on the morphological features, corrosion and/or biological properties, and the studies on nanomechanical and adhesion properties of MAO coatings have been seldom reported [26]. What is more, many reports have been published on the micro-arc oxidation of titanium and its various alloys, but only little of these concern the MAO process of the very promising alloy Ti13Zr13Nb, which is characterized by excellent biocompatibility, corrosion resistance, high strength/weight ratio, good fatigue resistance, and lower Young's modulus as compared to most of the metals [27]. Additionally, there is still very little reports on the micro-arc oxidation coatings deposited on the selective-laser melted (SLM) substrate [28]. Designing personalized implants can improve their longevity. Among others, selective laser melting (SLM) shows several advantages over conventional manufacturing techniques as a method for customized mechanical properties [29,30]. Some authors indicate that the high thermal gradient and high solidification of SLM-made alloys may affect MAO coating characteristics. For example, Yao et al. [28] obtained higher porosity with relatively smaller pores on the surface of the MAO coating applied on TC4 alloy produced by SLM, compared to other works. Some authors explained this phenomenon by an appearance of many small micro-arc oxidation discharges related to small grains size obtained by the SLM manufacturing process [31]. There are also studies, which indicate that the SLM/MAO process improves the osseointegration capacity [32,33]; however, the improvement of surface properties is rather related to the ability of additive manufacturing to produce the 3D structure which optimizes the contact interface with a human bone, than with a special microstructure characteristic.

To our best knowledge, there are no reports of the complex correlations between the MAO process parameters and growth mechanism, structural characteristics, adhesion strength, and nanomechanical behavior of MAO coatings on the selective-laser melted β-rich Ti13Zr13Nb alloy.

The main aim of this study was to investigate the complex effect of micro-arc oxidation process parameters (applied voltage, current, and duration time) in Ca- and P-containing electrolyte on the surface characteristics, early-stage bioactivity, nanomechanical properties, and adhesion between the MAO coatings and SLM-made Ti13Zr13Nb alloy. The MAO process was performed on the SLM-made and mechanically polished Ti13Zr13Nb alloy. Specimens were treated in 0.1 mol·L^{-1} of calcium glycerophosphate $C_3H_7CaO_6P$ (GP) and 0.15 mol·L^{-1} of calcium acetate $Ca(CH_3COO)_2$ (CA) by the DC power supply under various voltages 200, 300, and 400 V and a constant current of 32 mA for 10 and 15 min, and a constant current of 50 mA for 10 min. In order to obtain the characteristics of coatings, the topography, surface roughness, pore diameter, elemental composition, crystal structure, and surface wettability were evaluated. The ability of calcium phosphate formation on oxide coatings was examined to obtain the bioactivity characterization. Hardness, maximum indent depth, Young's modulus, and $E_{coating}/E_{substrate}$, H/E, and H^3/E^2 ratios were determined with the nanomechanical evaluation while the MAO coating adhesion was estimated by the scratch test.

2. Materials and Methods

2.1. Specimens Preparation

The cylindrical specimens (discs 20 mm in diameter and 3 mm thick) were manufactured by the selective laser melting (SLM) additive technique while using Ti13Zr13Nb spherical powder

(TLS Technik GmbH & Co. Spezialpulver KG, Bitterfeld-Wolfen, Germany) with particle size ranging from 20 to 70 μm. Samples were designed using the Materialise Magics (Materialise NV, Ghent, Belgium) software. The used SLM 100 apparatus (Realizer GmbH, Borchen, Germany) was equipped with the ytterbium one mode fiber laser CW YLR-100-SM (IPG Laser GmbH, Burbach, Germany) using a 1070 nm wavelength. The laser melting process was carried under a protective argon atmosphere and other process parameters (i.e., spot diameter, laser power and scanning speed, and layer thickness) are patent pending. The samples were mechanically polished using #150, #320, #600, and #800 grid SiC abrasive papers and ultrasonically cleaned with acetone, isopropanol, and distilled water for 10 min each. After ultrasonication, the specimens were dried in ambient air.

2.2. MAO Coating Preparation

The micro-arc oxidation process was performed with a DC power supply (PL-650-0.1, Matsusada Precision Inc., Shiga, Japan) under various voltages 200, 300, and 400 V and a constant current of 32 mA for 10 and 15 min, and a constant current of 50 mA for 10 min. The labels of MAO coatings formed at the different parameters are shown in Table 1. The MAO was conducted in 1 L of an aqueous electrolyte with contents of 0.1 mol·L^{-1} calcium glycerophosphate $C_3H_7CaO_6P$ (GP) and 0.15 mol·L^{-1} calcium acetate $Ca(CH_3COO)_2$ (CA), as P and Ca ions source, based on previous studies [34,35]. Ti13Zr13Nb was used as an anode and a cylinder made of AISI 304 stainless steel as a cathode. Each disk was fixed onto a polytetrafluoroethylene holder with an O-ring. Details of the working electrode were as described in [36]. A magnetic stirrer rotating at 200 rpm was used to create a turbulent flow regime, and the glass container was kept in a water-cooled bath at room temperature. After the MAO treatment, the surfaces were washed in ultrapure water and dried. The control group called "control" are represented by the Ti13Zr13Nb SLM-made specimens after polishing without the MAO treatment.

Table 1. The labels of micro-arc oxidation (MAO) coatings formed at different process conditions.

Applied Voltage (V)	Current (mA)	MAO-Treatment Time (min)	Lebels in Groups	Labels in Subgroups
200	32	15	MAO_200	MAO_32_15_200
	32	10		MAO_32_10_200
	50	10		MAO_50_10_200
300	32	15	MAO_300	MAO_32_15_300
	32	10		MAO_32_10_300
	50	10		MAO_50_10_300
400	32	15	MAO_400	MAO_32_15_400
	32	10		MAO_32_10_400
	50	10		MAO_50_10_400

2.3. MAO Coating Characterization

Process parameters (current and voltage) were continuously recorded at intervals of 0.1 s by a sensor interface (PCD-300A; Kyowa Electronic Instruments Co., Ltd., Tokyo, Japan). The content of the ripples was controlled to less than 0.1%. Scanning electron microscopy (SEM; S-3400NX, Hitachi High-Technologies Corp., Tokyo, Japan) was used to characterize the surface morphology. The samples' topography, surface roughness, and pore diameter, were evaluated by the laser scanning microscope (LSM; Olympus LEXT OLS4100 3D, Tokyo, Japan). The thickness of MAO coating was measured by the coating thickness gauges Elcometer 456 (Elcometer, 456, Elcometer Inc, Michigan, MI, USA) in a range of 0–1500 μm and ±1% accuracy. The elemental compositions of the samples were analyzed using energy dispersive X-ray spectrometry (EDS; S-3400NX, Hitachi High-Technologies Corp., Tokyo, Japan). X-ray diffraction (XRD, BRUKER D8 DISCOVER, Bruker AXS KK, Yokohama, Japan) was performed to characterize the crystal structure of the specimens. The surface wettability was determined by the contact angle (CA) measurements with the falling drop method while using an optical tensiometer (Attention Theta Life, Biolin Scientific, Espoo, Finland). The volume of the liquids was about 1 µL/sample. Measurement was carried out immediately after the deposition of the drop of the liquid while using the OneAttension program (Biolin Scientific, Espoo, Finland).

2.4. Bioactivity of the MAO Coatings

The ability of calcium phosphate formation on the oxide coatings was evaluated by an immersion test in Hank's solution (0.005 L) with a pH value of 7.4 and concentration similar to the extracellular fluid. The composition of the fluid is shown in Table 2. The specimens were immersed in the solution at 310 K for 72 h. After the test, specimens were rinsed in ultrapure water. Calcium phosphate formation on the specimens was evaluated with the scanning electron microscopy (SEM; S-3400NX, Hitachi High-Technologies Corp., Tokyo, Japan) and energy dispersive X-ray spectrometry (EDS; S-3400NX, Hitachi High-Technologies Corp., Tokyo, Japan).

Table 2. Ion concentration of Hank's solution.

Ion	Concentration (mol·L^{-1})
Na$^+$	1.42×10^{-1}
K$^+$	5.81×10^{-3}
Mg^{2+}	8.11×10^{-4}
Ca^{2+}	1.26×10^{-3}
Cl$^-$	1.45×10^{-1}
PO$_4^{3-}$	7.78×10^{-4}
SO$_4^{3-}$	8.11×10^{-4}
CO$_3^{2-}$	4.17×10^{-3}

2.5. Nanomechanical Properties of MAO Coatings

Nanomechanical properties of MAO coatings, i.e., hardness, maximum indent depth, Young's modulus, and $E_{coating}/E_{substrate}$, H/E, H^3/E^2 ratios were determined using a nanoindenter (NanoTest Vantage, Micro Materials Ltd., Wrexham, UK) with Berkovich three-sided pyramidal diamond with an apex angle equal to 124.4°. The tests were performed with a maximum force of 50 mN, loading and unloading times equaled to 20 and 15 s, respectively, and the cycle had 5 s dwell at a maximum load. Hardness (H), reduced Young's modulus (E_r), and Young's modulus values were determined using the Oliver-Pharr method [37] based on the NanoTest results analysis program. In order to convert the reduced Young's modulus into Young's modulus, a Poisson coefficient of 0.3 was assumed for the coatings [38]. For all tested specimens, the twenty-five independent measurements were performed.

2.6. MAO Coatings Adhesion Properties

The adhesion strength of the MAO coatings was estimated by NanoTest™ Vantage (Micro Materials, Wrexham, UK) using the Berkovich three-sided pyramidal diamond the same as mentioned above. The test parameters were as follows: Load equal was 400 mN, the loading rate was 1.3 mN/s, and scratch length was 10 µm. The way of loading was linear and continuous. The adhesion of the coating was assessed based on the observation of an abrupt change in frictional force during the test. The failure events were examined by an optical microscope.

2.7. Statistical Analysis

Statistical analysis of the data was performed with the commercial software OriginPro 64 (OriginLab, Northampton, MA, USA). The Shapiro-Wilk test was used to assess the normal distribution of the data. All of the results were presented as the mean ± standard deviation (SD), and they were statistically analyzed with a one-way analysis of variance (one-way ANOVA). Multiple comparisons versus the control group between means were performed using the Bonferroni t-test with the statistical significance set at $p < 0.05$.

3. Results and Discussion

3.1. MAO Process Characterization

The MAO process characterization was analyzed by the time-dependent relations of voltage and current, as shown in Figure 1. The curve characteristic is typical for the MAO process, where the variation of voltage can be divided into three stages [39]. First, the voltage increases linearly due to the formation of an initial dioxide film. During the second stage, the voltage increases slowly, which corresponds to the beginning of the micro-arc oxidation process, where the oxide layer is broken down to form a discharge channel. As the reaction progresses, with the increase of layer thickness, the voltage keeps increasing. During the last stage of the MAO process, the voltage fluctuates near the plateau with numerous sparks on each sample surface until the end of the treatment. A similar behavior was also observed by Qian [40]. Similarly, during analyzing the current-to-time curve, we can notice that the current decreases with increasing the MAO treatment time. In the first stage, the current is constant and stays on its highest value due to the formation of a compact and thin oxide layer (resulting in a better surface conductivity). When the system reaches the selected potential, a dielectric breakdown of these films and micro-arc discharge start to occur. With layer growing, the conductivity decreases, the sample electric resistance increases, and the current density decreases intensively [12]. Finally, the thickness of a coating prevents dielectric breakdown, and the current takes the minimum values near a plateau section of the current. This behavior is confirmed by the reports of Sobolev et al. [41] and Komarova et al. [21], and is correlated with optical observations. During the first stage, gas bubbles on the materials' surface could be seen. During the second stage, white and tiny electric sparks appear, and with increasing time, they become larger and more intense. In the third stage, the MAO process varies in different conditions. When the MAO treatment was conducted at 400 V, the arcs became stronger while their number decreased. In contrast, when oxidation was performed at 200 V, after reaching the maximum voltage, the micro-arc sparks became weak and fewer in number. It can be seen that the intensity of the micro-arc oxidation increases while the value of maximum voltage increases. The presence of pulses on the curves indicates the periodic nature of electrical discharges. The highest dielectric breakdown behavior of the process at an applied maximum voltage of 400 V could be observed.

From a comparison of curves, it can be noticed that at current 32 mA and time 15 min, the maximum voltage of 200, 300, and 400 V is reached after 80, 261, and 641 s, respectively (the accuracy of determining the time was estimated at 2 s and current value at 0.001 A. When the time of the MAO process was reduced from 15 to 10 min, the relations of MAO_200 and MAO_300 were similar to those observed for 15 min (Figure 1a,a' and b,b'), while for MAO_400, the limited time of 10 min occurred to be too short to obtain an expected voltage (the maximum registered value was 391 V). The incomplete MAO process significantly affected the coatings' characteristics. The time needed to obtain the final voltage 200, 300, and 400 V was about two-times shorter (42, 153, and 381 s, respectively), when the initial current increased from 32 to 50 mA (Figure 1b,b' and c,c') indicating that the film could be faster formed at these process parameters. The average value of current obtained in the third stage of the process was similar under voltage 200 and 300 V, but if the treatment voltage was 400 V, the average minimum values near a plateau section were higher compared to those under 200 and 300 V. The effect can be attributed to the changing electrical resistance of the oxide layer.

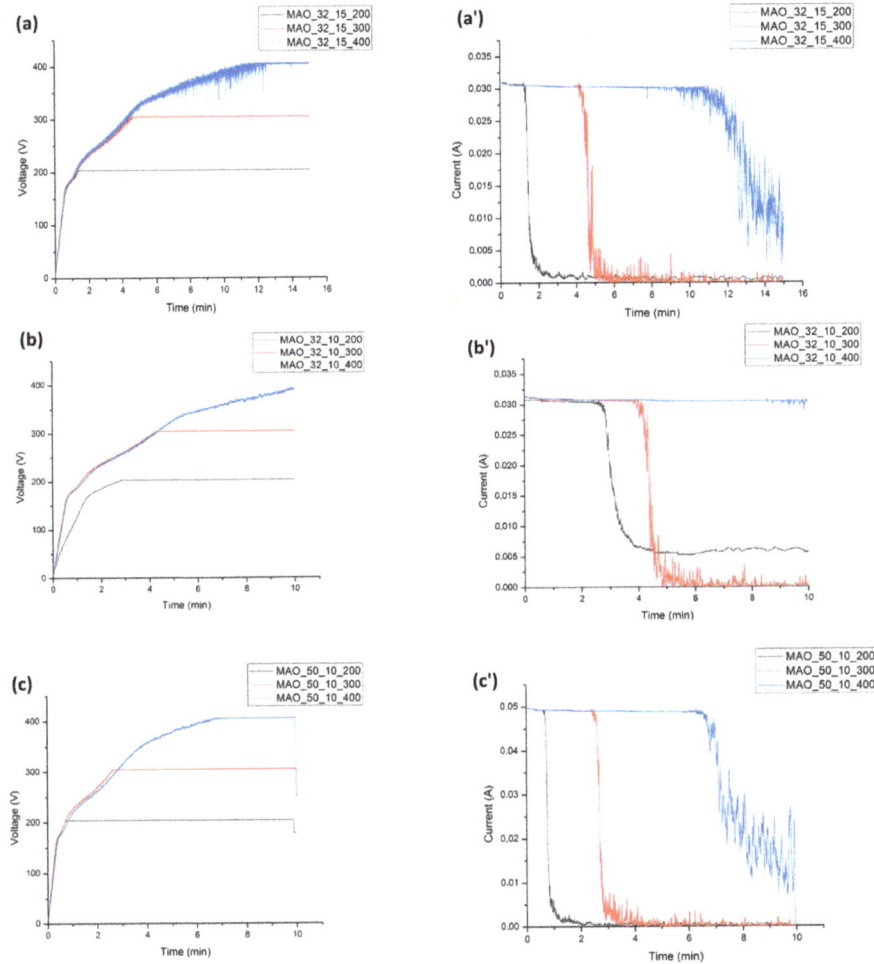

Figure 1. Time-dependent curves of voltage (**a**–**c**) and current (**a'**–**c'**) of the MAO process for MAO_32_15 (**a,a'**), MAO_32_10 (**b,b'**), and MAO_50_10 (**c,c'**).

3.2. MAO Coating Surface Evaluation

The surface morphology of the Ti13Zr13Nb alloy samples treated by the MAO in various process conditions was observed by SEM, as shown in Figure 2, the dimensions of the pores were calculated using ImageJ software, while the MAO coating thickness was evaluated by using coating thickness gauges. The average diameter of pores varied from 1.8 to 8.4 μm (Figure 3a), with applied voltage increasing in the range of 200–400 V, for all current and time values. Samples treated under lower maximum voltages, 200 and 300 V, possessed similar average pore sizes, however, their shape and distribution differed from each other significantly. At the lowest voltage (MAO_200), the samples' surfaces revealed mainly few irregularly distributed micropores with size equal to 2.17 ± 0.89, 2.20 ± 0.66, and 1.80 ± 0.64 μm (for MAO_32_15, MAO_32_10, and MAO_50_10, respectively). Additionally, on the surface, two characteristic regions can be observed (Figure 2a) with micropores (region II) and without micropores (region I). It could be assumed that at the selected potential 200 V, a dielectric breakdown of films and micro-arc discharge was not effective to form a fully porous coating. When limited voltage was 300 V, the sustainable and stable (Figure 1b) electrolytic plasma discharge resulted in regularly distributed pores over the surface with a relatively uniform pore diameter 2.36 ± 0.4, 2.80 ± 0.67, and 2.75 ± 0.61 μm for MAO_32_15_300, MAO_32_10_300, and MAO_50_10_300, respectively. As the MAO was performed at 400 V, the surfaces showed a crater-like structure with statistically the biggest pores about 7.13 ± 2.68, 5.66 ± 1.38, and 8.41 ± 3.19 μm for MAO_32_15_400, MAO_32_10_400, and MAO_50_10_400, respectively. The size of the micropores increased as some discharge channels connected and became larger. What is more, in these process conditions, the presence of tiny cracks can be observed (Figure 2c). The formation of microcracks may be caused by thermal stress [42] either due to the breakdown of the electrolytic plasma bubbles at the alloy surface which constantly release a large amount of energy, by rapid condensation of the molten compound during their exposition to the cold electrolyte [43] or rapid oxides' transformation from amorphous to crystalline form [44]. In a similar study, Correa et al. [14] observed qualitatively similar MAO samples' morphology of the Ti15ZrxMo alloy subject to the anodic treatment carried out for 10 min at the same electrolyte concentration (Ca + GPa) under the maximum voltage limited to 300, 350, and 400 V. In this case, the pore sizes were smaller as compared to our study and were as follows: 1.6–1.8 μm for 300 V, 2.2–2.8 μm for 350 V, and 2.8–4.5 μm for 400 V. The addition of MO which improves the corrosion resistance may have distinct effects on the dielectric barrier of the films.

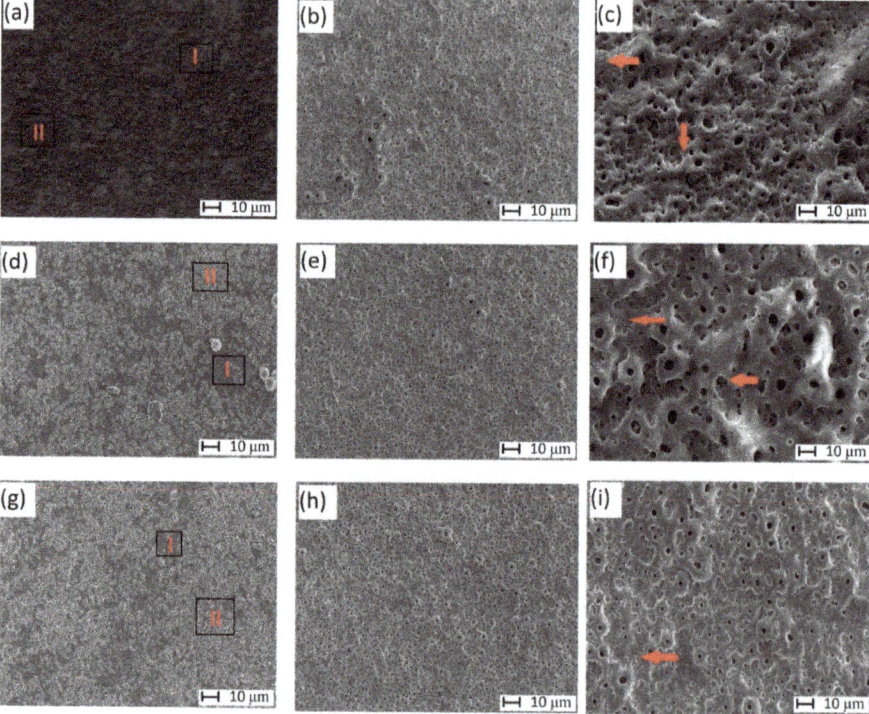

Figure 2. SEM microstructure images of the MAO coatings with various MAO process parameters: MAO_32_15_200 (**a**), MAO_32_15_300 (**b**), MAO_32_15_400 (**c**), MAO_32_10_200 (**d**), MAO_32_10_300 (**e**), MAO_32_10_400 (**f**), MAO_50_10_200 (**g**), MAO_50_10_300 (**h**), MAO_50_10_400 (**i**) with two various marked regions (I, II) obtained in MAO_200 (**a**) and microcracks obtained in MAO_400 (**c**) (the presented results are representative for three analyses of each surface treatment specimens).

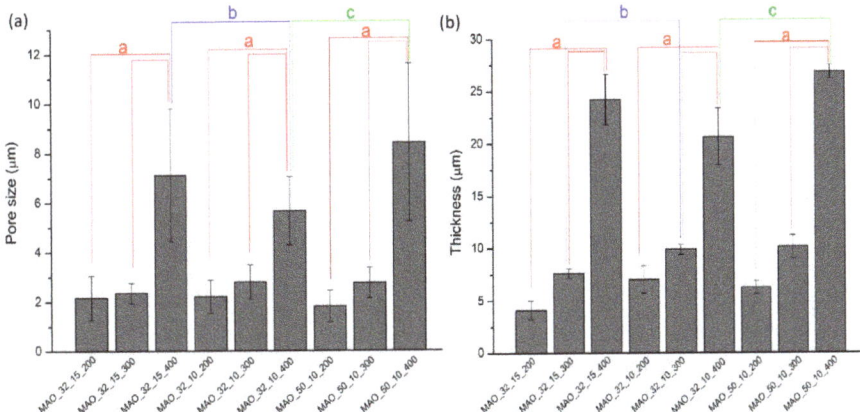

Figure 3. Pore size (**a**) and thickness (**b**) of the investigated MAO coatings with various MAO process parameters measured by ImageJ software and coating thickness guages, respectively (n = 25; data are the means ± SD; (a) significantly different in groups ($p < 0.05$), (b) significantly different between time groups ($p < 0.05$), (c) significantly different between current groups ($p < 0.05$)).

Taking into consideration the standard deviation value of pore size, the most extensive and hardly reproducible structure was obtained at the highest maximum voltage, especially at the highest current parameter for MAO_50_10_400. While, the most regular and predictable structure was obtained for the middle value of voltage (300 V), in particular for MAO_32_15_300. It can also be noticed that in the case of MAO coatings treated in 400 V, shortening the process time from 15 to 10 min in the same current conditions, decreases the pore size, while increasing the current from 32 to 50 mA in the same time process conditions results in a bigger pore size (Figure 3).

The anodic treatment resulted in pores with a wide range of sizes. Alves et al. [12], applying the MAO at similar anodic treatment conditions, obtained smaller pore sizes (0.5–5 μm), compared to our results. The decrease of the pore size could be caused by lower electrolyte concentration (GP and CA). The increase in the size of the pores with increasing electrolyte concentration was also reported by Ishizawa et al. [45] for anodic titanium oxide films containing Ca and P.

The MAO coating thickness also increased with increasing voltage and varied from 4.1 to 26.8 μm, for MAO_200 and MAO_400, respectively. Sedelnikova et al. [16] reported that CaP coatings on Ti and Zr1Nb increased linearly from 10 to 110 μm with the voltage rising from 150 to 300 V. Chen et al. [46] obtained the MAO film on Ti39Nb6Zr of 15 μm in thickness during a 10 min treatment at 400 V. While Karbowniczek et al. [47] observed that the MAO coatings ranged from 16 to 60 μm in thickness, depending on the Ca/P ratio in an electrolyte. According to the literature, when a current density grows, the coatings deposition rate increases [48]. However, in our study only MAO_400 fulfills this assumption.

The results of microstructural characterization are in line with the average surface roughness (Ra) where with increasing voltages, the Ra value also increases (Table 3). The significant ($p < 0.05$) highest value of Ra parameter was observed at 400 V in each time and current conditions. MAO_32_10_400 has a statistically lower Ra parameter compared to other samples treated in 400 V, due to the "unfinished" micro-arc oxidation process (Figure 1b). When the MAO process was carried out at the same time oxidation but under higher current, the Ra parameter increased, its value being close to that of MAO_32_15_400. The obtained results are consistent also with SEM images, in which better surface development is observed at increasing voltage. Increasing values of roughness parameters for MAO_400 are associated with the existence of the additional craters and a hill-like structure caused by aggressive plasma discharge. The obtained results are in line with previous reports [16,25,49]. Karbowniczek et al. [47] using the same electrolyte but under pulsed DC with 400 V and time of 5 min obtained Ra = 1.6 µm. Lin et al. [46] reported on Ra = 1.86 µm for the MAO treated Ti39Nb6Zr alloy after 10 min under 400 V. Sedelnikova et al. [16] showed that alloys containing Zr and Nb demonstrated lower values of roughness compared to pure Ti.

Table 3. Average surface roughness Ra (µm) of the investigated control specimens and MAO coatings with various MAO process parameters measured by laser microscopy.

Sample	Average Surface Roughness Ra (µm)			Control
	200 V	300 V	400 V	
MAO_32_15	0.41 ± 0.05	0.52 ± 0.01	2.67 ± 0.25 *,a	
MAO_32_10	0.37 ± 0.03	0.54 ± 0.26	1.84 ± 0.17 *,a,b	0.19 ± 0.04
MAO_50_10	0.36 ± 0.02	0.51 ± 0.03	2.67 ± 0.18 *,a	

(n = 5; data are the means ±SD; * significantly different from control ($p < 0.05$); [a] significantly different from MAO_200 and MAO_300 ($p < 0.05$); [b] significantly different from MAO_400 ($p < 0.05$)).

The EDS microanalysis of the MAO coatings revealed the different amounts of elements throughout the coating surface in relation to the MAO process conditions (Figure 4). The MAO coatings formed in the electrolytes are mainly composed of Ti and O. The oxygen concentration on MAO samples is about 60%, which is the result of intense anodization [50]. As seen, in each of the MAO specimens, Ca, P, Ti, Zr, and Nb elements appear. It could be speculated that Ca and P from the electrolyte were incorporated into the oxide layers on the substrates during the arc discharge process, while the Ti, Zr, and Nb elements were detected from the titanium alloy substrate [51]. It can be also noticed that with the increasing applied voltage, the concentration of Ca increases, Ti decreases, while the P compound content first intensively increases with the increasing voltage ranging from 200 to 300 V and then slightly decreases. This phenomenon has been previously observed [52] and can be correlated with an increase in the intensity and temperature of the micro-arc discharges leading to the increase of the reactive capacity of all electrolyte components. It was also found [45] that sodium glycerophosphate and calcium acetate were suitable for the electrolytes to form a coating possessing the Ca/P ratio as that of hydroxyapatite (1.67). In this study, with the increasing voltage, the Ca/P ratio also increased; however, the value was smaller than that of bone-apatite. The highest value was measured for MAO_32_15_400 while the lowest for MAO_32_15_200 specimens (1.38 and 0.31, respectively). On the other hand, Han et al. [22] presented a study showing that specimens with a Ca/P atomic ratio of 1.73 did not exhibit apatite-induced ability, while the specimens with higher Ca and P content, but lower Ca/P atomic ratio induced an apatite appearance on the surface. These observations confirm that the MAO process not only affects the surface morphology but also influences the chemical composition of the oxide films as well.

Additionally, the high value of carbon (C) content, especially for control specimens, can be noticed. According to the literature, it is very difficult to obtain and analyze the reliable value of C content using energy dispersive X-ray spectrometry and the results significantly vary from each other from 2.1–5.7 (wt.%) [6], even to more than 27 (wt.%) [24], depending on the research group. Some authors did not include in their analysis the C content [51] at all. The high value of carbon detected during the EDS examination may be related to the contamination coming from the CO_2 atmosphere during the MAO treatment, from the internal walls of the vacuum chamber in SEM, as well as from accidentally improper samples storage [53]. In the future, it is worth considering eliminating the EDS analyses of such elements such as carbon.

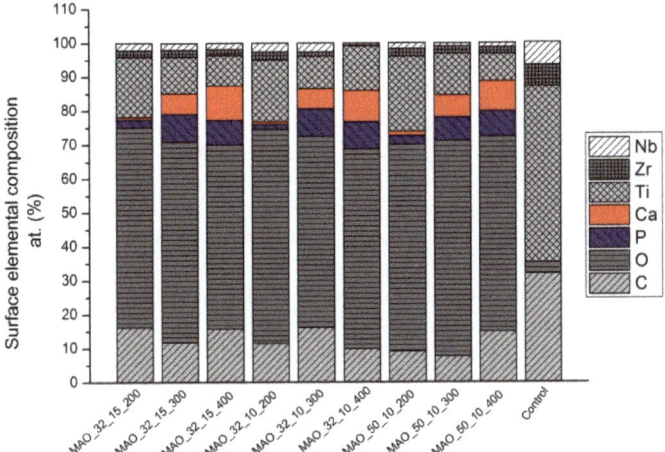

Figure 4. Surface elemental composition analysis (at.%) of the investigated control specimens and MAO coatings with various MAO process parameters measured by energy dispersive X-ray spectrometry EDS (the presented results are representative for three analyses of each surface treatment specimens).

In Table 4, the wettability measurements of each specimen are presented. They revealed that all specimens were characterized by hydrophilic surfaces (CA < 90°). The contact angle of the substrate and MAO coatings were in the range 35–64°. As a comparison, Ming-Tzu Tsai [54] obtained for the MAO treated Ti specimens, the contact angle equaled to 50.7° ± 5.6°. The wettability of a surface is crucial for the adsorption of proteins and the cell adhesion [38], with optimal value suggested for 35–85° [55]. Although only the MAO_32_10_400, MAO_50_10_400, MAO_32_10_200, and MAO_32_15_300 specimens significantly ($p > 0{,}05$) improved the surface wettability compared to the control group; it could be said that all specimens fulfilled this criterion. The literature [14] discloses that the contact angle of the MAO coatings increases with the rising applied voltages, which may be correlated (among others) with a higher value of roughness [19], and/or capillary forces between volcano-like pores and contacted distilled water [56]. However, in this study, only MAO_32_15_400 is in line with this relation, but surface wettability may be strongly affected by other parameters such as morphology, crystallinity, and chemistry [57].

Table 4. Surface wettability of the investigated control specimens and MAO coatings with various MAO process parameters determined by the measurements of the contact angle of distilled water.

Sample	Surface Wettability—The Value of Contact Angle (°)			Control
	200 V	300 V	400 V	
MAO_32_15	56.65 ± 6.98	35.63 ± 4.35 *,b	64.31 ± 7.94	
MAO_32_10	53.99 ± 2.56 *	57.40 ± 1.8	39.16 ± 5.18 *,a	60.83 ± 3.34
MAO_50_10	64.82 ± 4.95 d	62.13 ± 2.58	46.55 ± 3.57 *,c	

(n = 5; data are the means ±SD; * significantly different from control ($p < 0.05$); [a] significantly different from MAO_32_10_200 and _300, MAO_32_15_400 ($p < 0.05$); [b] significantly different from MAO_32_15_200 and _400; MAO_32_10_300 ($p < 0.05$), [c] significantly different from MAO_50_10_200 and _300 ($p < 0.05$) [d] significantly different from MAO_32_10_200 ($p < 0.05$)).

The XRD patterns of a control group and the micro-arc oxidized samples obtained at various applied voltages, and constant current 32 mA and time processing of 15 min are shown in Figure 5. In a case of a control group (Figure 5d), besides titanium peaks (α-, β-phase) related to the titanium alloy substrate, the rutile and anatase oxides were detected. It confirms a good ability of Ti13Zr13Nb alloy to self-passivation [58]. On the surface after the MAO treatment, among the titanium oxides, rutile, and anatase peaks appeared at all the applied voltages (Figure 5a–d). Additionally, at the lowest values of voltage (200 and 300 V), titanium peaks (α-phase) were also observed (Figure 5a,b). The intensity of the peaks slightly decreased with the increasing limiting voltage, which could be related to the oxide layer growth. For the highest applied voltage (400 V) (Figure 5c), the oxidized surface layer became mainly a mixture of anatase and rutile. The presence of anatase suggests that a vigorous oxidation reaction has occurred on a titanium surface during the MAO process [2]. The Ti peak on the XRD patterns may be correlated with X-ray penetration through the coating. According to the previous study, TiO_2 can exist in the three crystalline polymorphs (anatase, rutile, and brookite). The literature also indicates that mostly anatase is formed at lower forming voltages, while the combination of anatase and rutile phases appear at higher forming voltages since anatase as a metastable-phase gradually transforms into rutile at higher temperatures with the increasing dielectric breakdown phenomena [59]. Wang et al. [25] showed that alloys contained Nb ad Zr elements could form a number of anatase phase grains higher than those of rutile phase during high voltage (400 V) oxidation, compared to pure Ti and Ti6Al4V, which enhanced their bioactivity. No more phases were detected by the XRD analysis despite the fact that the EDS analysis also showed incorporation of Ca and P ions. The additional phases may appear in the amorphous or crystalline structure at a content that is below the detection limit of XRD [22]. These phenomena are in line with the literature; Zhang et al. [60] did not find any peaks from hydroxyapatite with the Ti-MAO sample in the CA and GP solution, while Han et al. [22] observed only Ca- and P-containing phase by employing the CA- and β-GP solution at higher voltages, up to 450 V.

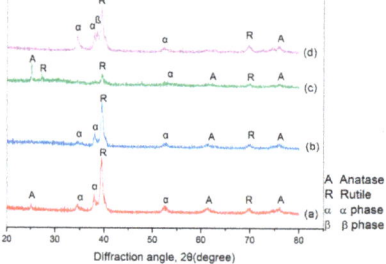

Figure 5. X-ray diffractograms of samples: The control group (**d**) and MAO coatings obtained at various applied voltages in a constant current of 32 mA and time processing equal to 15 min: MAO_32_15_200 (**a**), MAO_32_15_300 (**b**), MAO_32_15_400 (**c**) (the presented results are representative for three analyses of each surface treatment specimens).

3.3. Early-Stage Bioactivity of the MAO Coating

To investigate the early-stage bioactivity of the substrate and MAO coating under various maximum voltages, the specimens were immersed in Hank's solution for 72 h. The surface morphologies of the example of the samples after immersion are shown in Figure 6. It can be noticed that there are a few white particles randomly scattered on the sample surface (Figure 6a,b) or riddle-like small particles (Figure 6c,d) rich in Ca and P elements (Figure 6e,f). Other authors also observed similar morphologies with the spherical particles and precipitate layer being reported as apatite [19,60,61].

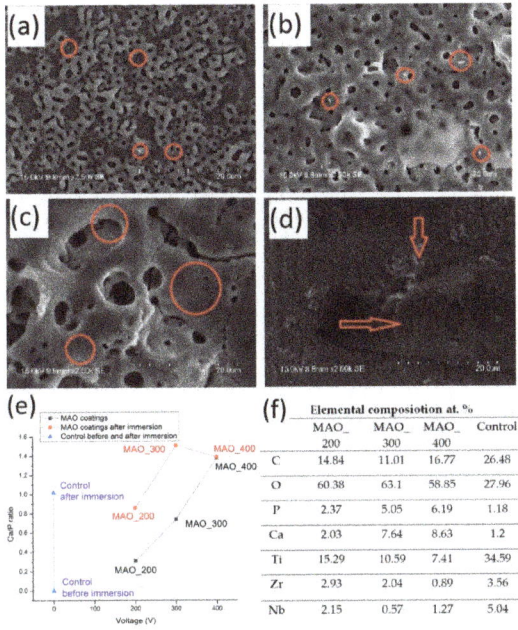

Figure 6. Surface morphologies of the samples after soaking in Hank's solutions for 72 h: The control group (**d**) and the MAO coatings obtained at various applied voltages in a constant current of 32 mA and time processing equal to 15 min (**a**–**c**): MAO_32_15_200 (**a**), MAO_32_15_300 (**b**), and MAO_32_15_400 (**c**). The Ca/P ratio (**e**) and elemental composition (**f**) of the post-immersed specimens (the presented results are representative for three analyses of each surface treatment specimens).

The elemental composition of the post-immersed substrate and samples with MAO coatings, and Ca/P ratio are shown in Figure 6e,f. It can be observed that with the increasing voltage, the Ca and P values also increase. Additionally, all post-immersed MAO specimens are characterized by a higher Ca/P ratio as compared to a substrate (control). The Ca/P ratio after immersion for specimens treated at 200 and 300 V is about two times higher than before immersion (0.85 and 1.51, respectively). While for specimens treated under 400 V, the Ca/P ratio decreases and is almost the same as the value before immersion. It is difficult now to propose any model explaining the relative decrease in P content with the increasing voltage, but it may be somewhat attributed to the different chemical potentials of deposition of both ions and their different relations of chemical potentials on the electrochemical voltage. The results are in line with a report of Qing et al. [19] that presents that the MAO coating formed at 350 V shows better apatite-inducing ability as compared to the MAO process carried out at higher voltages (400, 450, and 500 V). The value of the Ca/P ratio being the nearest to stoichiometric hydroxyapatite (1.67) was obtained by specimens treated under 300 V (1.51). Anyway, the ratio Ca/P demonstrates the surfaces' ability to induce the precipitation of calcium phosphate.

According to the literature, due to nucleation of apatite and diffusion of Ca and P ions from the MAO coating into Hank's solution, the degree of local supersaturation increases, and apatite nuclei are spontaneously growing. Moreover, the increase in the aperture and surface roughness of the MAO coating also accelerate the nucleation of apatite [28]. Thus, the bioactivity of MAO increased with a voltage change from 200 to 300 V. Its further decrease can be, as mentioned above, due to many factors, including relation of chemical potential on the voltage. In a similar study, Tsutsumi et al. [23] observed on MAO-treated Ti29Nb13Ta4.6Zr the newly formed layers rich in Ca and P after seven days of immersion in Hank's solution. Correa et al. [14] presented that the calcium and phosphate contents increased with the limiting voltage in all samples after immersion. Our results, confirmed also by other research reports [19], indicate only that at high voltage the enhancement of bioactivity may decrease.

3.4. Nanomechanical Properties of MAO Coatings

The load-displacement hysteresis curve as an example of the results of nanoindentation tests is shown in Figure 7. Three main stages of the nanoindentation test are observed: Loading to a maximum value, pause with maximum value, and unloading. At the end of the loading stage, the maximum indenter displacement connected with elastic and plastic deformation can be observed, while at the end of the unloading phase, the final depth (connected with elastic recovery) of the contact impression can be the measure [37]. Irregularity occurring on the unloading step is correlated with the temperature drift. It can be seen that the elastic recovery of MAO specimens is significantly low when compared to control specimens and decreases with an increasing voltage of the MAO process. The lowest recovery characteristic was obtained by the MAO_32_10_400 treatment, while the greater by MAO_32_10_200. Increasing current values, from 32 to 50 mA, result in lower maximum displacement, while decreasing time of the MAO process from 15 to 10 min leads to higher maximum displacement. What is more, at the higher voltage of the MAO process, the longer displacement during pause time can be observed, which is connected with more distinct creep of these coatings. The results differ from the phenomena shown in the literature, where the MAO process significantly has improved elastic recovery compared to the uncoated substrate [62,63].

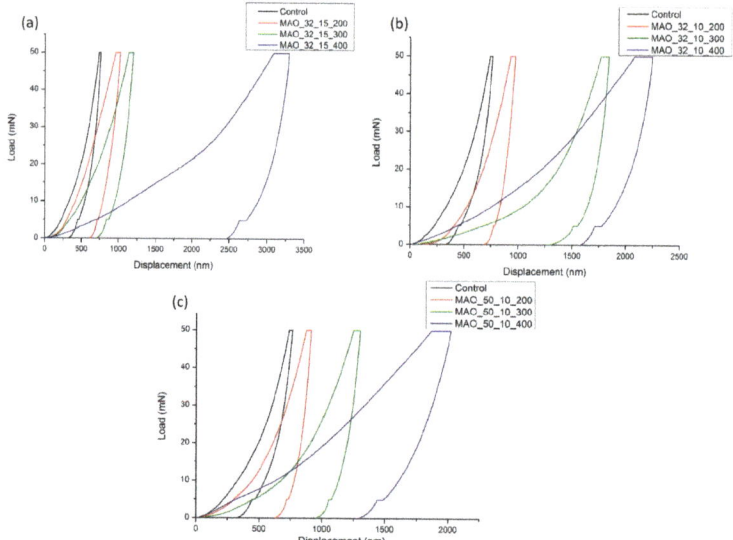

Figure 7. Nanoindentation load-displacement curve obtained for the investigated MAO coatings with various MAO process parameters and control specimens obtained by the nanoindentation test (**a**–**c**): MAO_32_15 (**a**), MAO_32_15 (**b**), and MAO_32_15 (**c**) (the presented results are representative for twenty-five analyses of each surface treatment specimens).

Hardness (H), Young's modulus (E) value, and deformation mechanism of the coating under load are considered as the most important factors for the anti-wear performance of ceramic coatings [64]. Figure 8 presents the surface hardness, Young's modulus, maximum indent depth, $E_{coating}/E_{substrate}$, H/E, and H^3/E^2 ratio of the MAO coatings. The highest value of hardness and Young's modulus was obtained for the control group (4.75 ± 1.55 and 67.71 ± 13.09 GPa, respectively). It can be noticed that all modifications had a significant impact on both mechanical indices. Apart from the

MAO_32_10_300 and MAO_32_10_400 specimens, with increasing voltages of the MAO process, the hardness significantly decreased, from two times for MAO_200 up to twelve times for MAO_400. In MAO_300 specimens, when the time decreased from 15 to 10 min, the hardness was reduced almost 60% compared to the longer process, while the increment of current from 32 to 50 mA caused almost twice higher hardness of tested coating compared to specimens treated at 32 mA. A similar behavior was noticed for Young's modulus where MAO_50_10_300 was characterized by a higher value of Young's modulus in comparison to MAO_32_10_300. In contrast, the time change from 15 to 10 min first decreased (for MAO_200) and then increased (for MAO_300) the Young's modulus value. A higher parameter of hardness in all cases was related to denser and smoother surface morphologies. Young's modulus was higher for all specimens treated under 200 V with the highest value equal to 80.94 ± 6.71 GPa obtained during the MAO process in 50 mA conditions, while specimens treated at 300 and 400 V showed the significantly decreased Young's modulus. The results are in line with the literature, where the lowest values of Young's modulus and hardness correspond to materials with less dense microstructure and the highest levels of porosity [65–68]. The MAO process carried out at 400 V indicated from two to four times bigger pore size compared to the MAO process at 200 V. Similarly, Souza et al. [69] suggested that the elastic modulus obtained for porous film was related to the actual elastic response of the film. In the case of biomaterials, lower values of Young's modulus are often positive. The mismatch between elastic modulus and hardness of an implant and a bone can lead to stress shielding, which may provide implant loosening [27,29]. The obtained results indicate that the elastic modulus values for all MAO coatings are closer to values characteristics of bones (11.4–21.2 GPa) and the best matching is observed for specimens treated at 400 V (8.46 ± 0.59; 14.06 ± 2.11; 15.00 ± 3.07 for MAO_32_15_400, MAO_32_10_400, and MAO_50_10_400, respectively). As presented in Figure 8b, with increasing voltage, maximum indentation depth also increases which is strictly correlated with lower hardness of the coatings. The highest indentation depth reached 2.55 µm (for MAO_32_15_400) and was about 60% higher than the lowest one equal to 0.94 µm (for MAO_50_10_200). According to the literature, the lower the H/E and H^3/E^2 ratios are, referred to materials elastic strain failure and materials resistance to plastic deformation, respectively, the worst the materials wear performance is [70]. It can be noticed that both ratios are much lower for the MAO coatings compared to the control group, and lower than the optimal declared value of the H/E parameter approx. 0.1 [71], which refers to high plastic deformation of the coating and in consequence, low wear resistance (Figure 8e,f). Some authors proved that a close match between the elastic modulus of the coating and substrate reduced interfacial stress and ensured better adhesion [72]. It is preferable to obtain the value of $E_{coating}/E_{substrate}$ ratio close to unity, since the higher mismatch, the greater plastic deformation and brittle cracking may be observed [73]. It can be noticed that $E_{coating}/E_{substrate}$ decreases with increasing applied voltage, and the closest value to the optimum is obtained by all specimens treated under lower voltage (MAO_200), while the highest mismatch between both coating and substrate's Young modulus occurs in the specimens treated under high voltage (MAO_400).

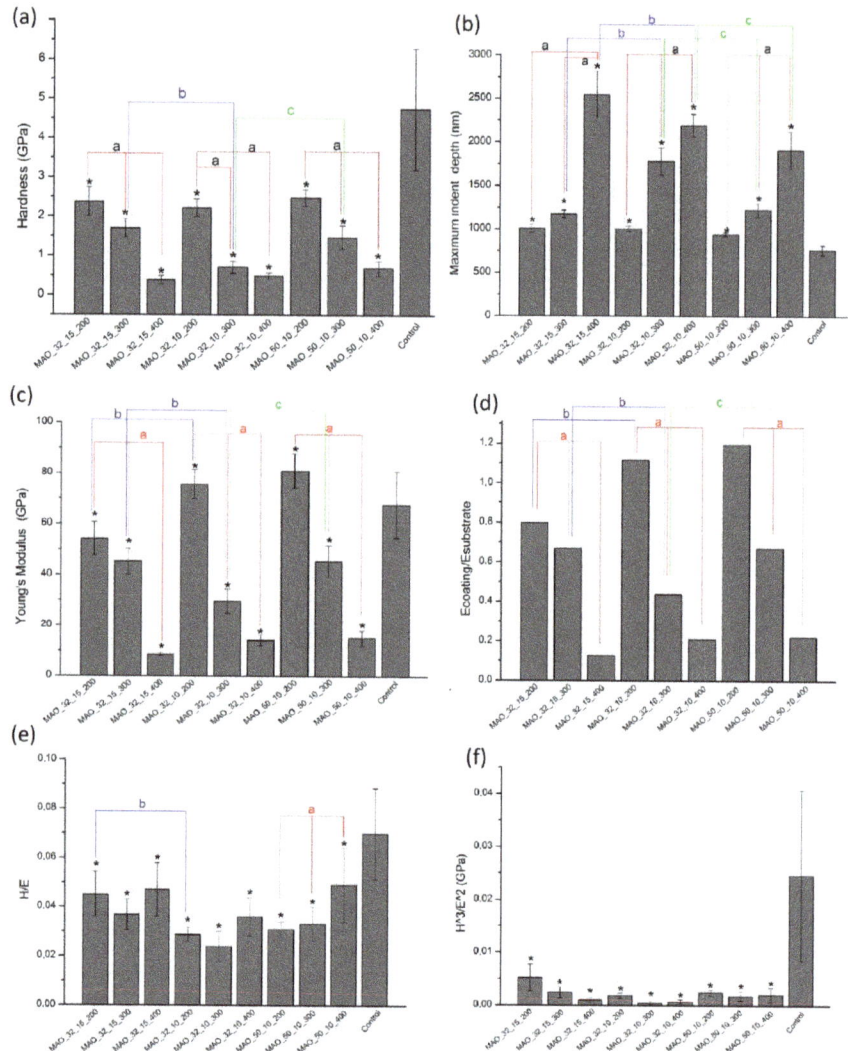

Figure 8. Nanomechanical properties: Nanohardness (**a**), maximum indent depth (**b**), Young's modulus (**c**), $E_{coating}/E_{substrate}$ ratio (**d**), H/E ratio (**e**), H^3/E^2 ratio (**f**), of investigated MAO coatings with various MAO process parameters and control specimens (n = 10; * significantly different from control ($p < 0.05$), (a) significantly different in groups ($p < 0.05$), (b) significantly different between time groups ($p < 0.05$), (c) significantly different between current groups ($p < 0.05$)).

According to the literature, mechanical parameters depend on many determinants, i.e., porosity, phase composition of the oxide layer, and MAO process parameters. It is suggested that soft calcium phosphate agglomerates have a significant impact on lower mechanical properties. For example, Karbowniczek et al. [47] obtained that the MAO coating containing various Ca/P ratios with hardness ranged between 5.74 to 21.88 GPa and Young's modulus from 29.3 ± 4.3 to 54.7 ± 11.8 GPa, in line with the results of this study. It is worth noting that the differences of the mechanical properties and their significant standard deviation values are characteristic of nanoindentation measurements as observed by many authors [74]. Especially, nanoindentation of macro-porous films is challenging due to the experimental errors related to the influence of the substrate, the pore-filling densification effect during indentation, as well as the surface roughness [75–77]. Additionally, the methodology to a reliable determining of the mechanical properties of porous coatings has not been well established yet. According to the literature, to minimize the effect of the substrate, the indenting should be provided to a maximum depth (h_{max}) less than 10% of the coating thickness (t_f) ($h_{max}/t_f < 0.1$) [37], and some authors indicate 20% of the film thickness for the porous thin ceramic films ($h_{max}/t_f < 0.2$) [78], while others pointed out that for hardness measurements, the substrate effect would be relatively small as long as the indentation depth would be below 50% of the film thickness [79]. On the other hand, taking into consideration the roughness effect, the elastic modulus results are more consistent and reproducible when the indentation depth is bigger, in the range of 10%–20% of the film thickness [65], since when the indenter tip is similar to the value of the surface roughness, the stresses at the initial point are much greater and result in variability in the first stages of indentation [80]. It can be noticed that in our research all MAO coatings fulfill the relations $h_{max}/t_f < 0.5$, while MAO_300 and 400 fulfill the relations $0.1 < h_{max}/t_f < 0.2$, which according to the literature should improve the reliability of the film-only properties. What is more, to eliminate the experimental errors, a series of indentation tests were carried out, the critical values were statistically rejected and the median was taken into consideration. Additionally, due to minimizing the pore-filling densification effect, the experiment was performed with restriction distance of 20 µm offset in the Z and Y axis. In the future, to improve the nanoindentation examination, the spherical indenter tip (with a larger surface area), as well as the nanoindentation tests evaluated at different depths should be considered.

3.5. MAO Coatings' Adhesion

In scratch testing, three main failures can be observed: Cohesive, adhesive, and mixed cracking [72]. The minimum load at which the first damage occurs is characterized by a critical load (Lc). In the literature usually one [74,81], two [82,83] or three [72,84] critical loads (Lc1, Lc2, Lc3) are presented. In this study, Lc1 is correlated with the first crack (initial cohesion deformation), Lc2 marked the load under which, on the edge of the scratch, the first delamination occurred (initial adhesion deformation), Lc3 presents total coating perforations, where the substrate starts to be seen continuously. According to the literature [85], Lc1 may be identified by the first peak of acoustic emission correlated with a spontaneous shock wave that occurs as a result of the microcrack formations, while Lc2 and Lc3 can be measured in the first and follow high-peaks which indicate the large damage of the coating. To improve the investigation of failures, all results were also analyzed while using other scratch test data, i.e., friction force and penetration depth. Since the literature indicated that isolated damages could be ignored [85], to understand better at which location, the beginning of the failure event was occurring, the optical microscopy was used.

A comparison of the microstructures of the coatings (Figure 9) shows that the scratch behavior is dependent on the MAO process parameters. For specimens treated under 200 V, only a small peeling occurred in the coating (apart of the place where the loading was completed, and bigger peeling could be observed), and only a few circular microcracks, some lateral and arc tensile cracks could be observed. The higher the process voltage, the more transverse semi-circular cracks, chipping, and spalling appear. For 300 and 400 V, conformal-type buckling cracks with local interfacial spallation indicate a highly cohesive failure. In both conditions at the edges of the scratches, two-sided peeling indicates the intense adhesive type of failure. However, the chipping deformations occur much later (farther from the beginning of the scratch path) compared to MAO_200 what may suggest the higher adhesion between substrate and coating. In specimens treated under 300 V conditions, total perforation becomes continuous, and in contrast, the coating's damage of specimens treated in 400 V presents more cyclic delamination character. Although in all specimens coating breakthrough occurs, the place where the substrate starts to appear along the scratch track, varies; the lower the voltage was, the earlier (closer to the beginning of the scratch path) the total perforations appeared. The example of the relations between friction (F), acoustic emission, penetration depth and load (L) with the indicated critical loads (Lc1,

Lc2, Lc3) for the MAO coatings are shown in Figure 9, while the average values of critical loads and critical friction force causing total delamination (Fc (Lc2)) measured during the scratch test for all MAO coatings samples are present in Table 5.

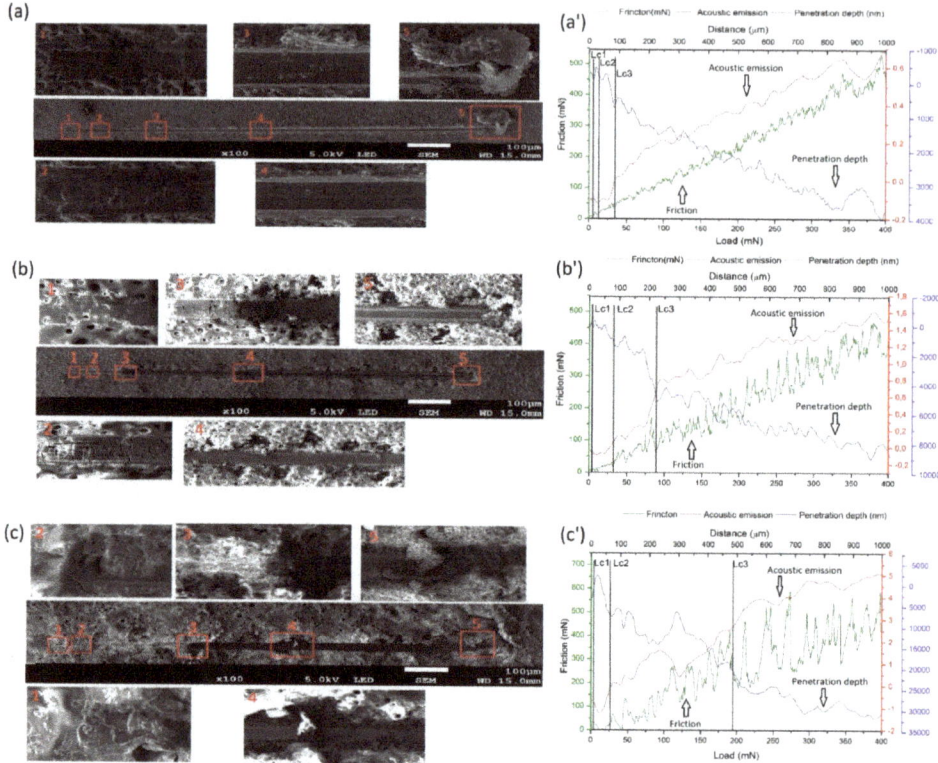

Figure 9. The scratch path with cohesion and adhesion failures after scratch testing (**a**–**c**), and the relation of friction (F), acoustic emission, and penetration depth on load (L) with the indicated critical loads (Lc1, Lc2, Lc3) (**a′**–**c′**) of the investigated MAO coatings obtained at various MAO process parameters: MAO_32_15_200 (**a**,**a′**), MAO_32_15_300 (**b**,**b′**), and MAO_32_15_400 (**c**,**c′**) (the presented results are representative for ten analyses of each surface treatment specimens).

Table 5. Critical loads and critical friction force caused total delamination (Fc (Lc2)) measured during the scratch test for all MAO coatings samples (the presented results are representative for five analyses of each surface treatment specimens).

	Critical Load (mN)			
Sample	Lc1	Lc2	Lc3	Fc (Lc2)
MAO_32_15_200	4.97 ± 1.47	12.34 ± 4.9	43.27 ± 8.16	220.68 ± 16.95
MAO_32_15_300	7.19 ± 4.17	27.41 ± 12.78	80.62 ± 11.39	134.48 ± 25.53
MAO_32_15_400	10.51 ± 1.68	63.49 ± 28.01	163.35 ± 42.13	368.332 ± 40.85
MAO_32_10_200	5.07 ± 1.94	29.70 ± 10.49	59.39 ± 9.56	39.064 ± 14.56
MAO_32_10_300	10.27 ± 3.63	31.28 ± 13.46	83.50 ± 8.06	45.14 ± 21.56
MAO_32_10_400	8.52 ± 2.26	46.63 ± 16.88	157.76 ± 36.69	138.42 ± 43.69
MAO_50_10_200	8.12 ± 1.89	17.63 ± 7.14	44.73 ± 4.26	131.80 ± 16.91
MAO_50_10_300	13.42 ± 3.93	25.56 ± 5.67	76.38 ± 6.98	477.37 ± 19.93
MAO_50_10_400	12.02 ± 3.99	43.13 ± 15.39	178.23 ± 43.81	520.08 ± 48

It can be said that for the MAO 200, 300, and 400 V conditions, the critical load of micro-cracking (Lc1) occurs at low force and ranges between 4.97 ± 1.47 to 13.42 ± 3.93 mN (for MAO_32_15_200 and MAO_50_10_300, respectively). According to Figure 9 and Table 5, the scratch resistance of the coatings was improved with the highest voltage of the MAO process, in line with the literature [62]. Some authors [28] suggested that the oxidation time influenced the adhesion; however, in this study, no significant difference could be observed. Lc3 of MAO_400 coatings (~157–178 mN) were approximately 50% higher than Lc3 of the MAO_300 and 70% higher than for the MAO_200. Additionally, MAO_400 remained without being totally damaged for the longest time (total delaminations occurring futher from the beginning of the scratch path) of all analyzed coatings (~370 μm for MAO_32_10_400 and MAO_32_15_400 and ~ 480 μm for MAO_50_10_400). In contrast, the fastest (the total delaminations occurring closer from the beginning of the scratch path) total delamination occurred in MAO_200 (~173, 85, and 101 μm for MAO_32_10_200, MAO_32_15_200, and MAO_50_10_200, respectively). In the specimens oxidized under middle voltages (MAO_300), the average distance from the beginning of the scratch path, where the first delamination occurred, were at ~205, 223, and 187 μm for MAO_32_10_300, MAO_32_15_300, and MAO_50_10_300, respectively. Even though MAO_400 is characterized by lower hardness and higher porosity compared to the MAO_200, better adhesion may be related to the higher P and Ca elements' incorporation due to a more intense oxidation process. The highest value of the friction force during the first adhesive damage Fc (Lc2) was obtained by the most rough specimens MAO_50_10_400 and MAO_32_15_400. The very high value of Fc (Lc2) of MAO_50_10_300 reflects the highest pill-up effect compared to other specimens (Figure 9b).

Although the MAO has been declared as an effective method to improve adhesion between coatings and substrate compared to other wide known surface modification methods, such as sol-gel [62], or alkaline heat treatment [86], a typical brittle behavior can be still observed. Based on the Zhu et al. [87] study, the tribological performance of the MAO coatings can be improved by an optimal combination of the MAO process parameters as voltage, current, time oxidation, or electrolyte concentration. It is also worth noting that results obtained in scratch tests depend on various independent factors (test conditions, material parameters, and the randomness of measurements), and their direct comparability between various research groups is almost impossible. Therefore, the obtained findings are only approximate and should be mainly used for the qualitative comparison purpose.

4. Conclusions

After the MAO treatment, rutile and anatase peaks appear at all the applied voltages. Mostly anatase is formed at lower forming voltages, while the combination of anatase and rutile phases appears at higher voltages. The time-dependent relations of voltage and current characteristics of the MAO process on SLM-made Ti13Zr13Nb are typical, their dielectric breakdown intensities increasing with the growth of applied voltage. The time 10 min is too short to obtain a maximum voltage of 400 V. The film is faster formed at higher initial current.

The average diameter of pores and thickness of the MAO coating depend on the applied voltage and change from 1.8 to 8.4 and 4.1 to 26.8 μm, respectively, with increasing voltage. For the highest voltage, shortening the processing time in the same current conditions, decreases the pore size, while increasing current results in a bigger pore size. The most roughness, extensive, and hardly reproducible structure is obtained at the highest maximum voltage, while the most regular and predictable structure at the middle voltage (300 V).

With increasing applied voltage, the concentration of Ca increases and that of Ti decreases, while the P compound content first intensively increases and then slightly decreases. With increasing voltage, the Ca/P ratio also increases. The closest value of the Ca/P ratio to bone-apatite is obtained by MAO_32_15_400.

All specimens have hydrophilic surfaces, with values corresponding to optimal values to ensure better adsorption of proteins and the cell adhesion.

After the early-stage bioactivity examination in Hank's solution, the MAO process results in a high Ca/P what indicates good bioactivity of MAO coatings. The nearest value of the CA/P ratio to stoichiometric hydroxyapatite is obtained by specimens treated at 300 V.

The elastic recovery of the MAO treated specimens is low and decreases with an increasing voltage of the MAO process. The significant elastic recovery appears at higher current values and higher processing time. Young's modulus is the highest for specimens treated at the lowest voltage 200 V. The elastic modulus values for all MAO coatings are closer to values characteristic of bones with the best matching for specimens treated at 400 V.

The ratios referred to coatings' wear-resistant performance (H, H^3/E^2, $E_{coating}/E_{substrate}$) show a tendency to higher plastic deformation of the MAO coating and in consequence, lower wear resistance compared to the uncoated Ti13Zr13Nb. The soft calcium phosphate agglomerates have a significant impact on lower mechanical properties. For the higher process voltage, more intense transverse semi-circular cracks, chipping, and spalling appear. The critical loads referred to the first adhesive damage and total delaminations are weaker at the highest voltage of the MAO process, which can be correlated with higher thickness of those coatings.

The results of the research indicate that there are no significant differences between MAO coating on the SLM substrate compared to MAO coating on titanium alloys produced by conventional methods. The differences in MAO coating properties were mainly correlated with micro-arc oxidation process parameters.

The most important parameter of the MAO process influencing the coating characteristic is the voltage. The optimal conditions of the MAO process include the voltage 300 V for 15 min at 32 mA or 50 mA of the current. As a consequence, the coating is characterized by the most regular and predictable structure, high Ca/P ratio, high hydrophilicity, highest demonstrated early-stage bioactivity, better nanomechanical properties, such elastic modulus and hardness with higher matching to the values of bones, compared to MAO_200 and uncoated substrate, as well as the higher critical force of adhesion and total delaminations compared to MAO_200. Despite the perspective properties, the relatively low critical load presented during the scratch test is still a certain problem; however, it is also an interesting opportunity to evaluate the issue of adhesion strength in further research. The authors plan to improve the adhesion properties, while adding some components (i.e., silica) to the one-step MAO process.

Author Contributions: Conceptualization, M.D., M.S., and A.Z.; methodology, M.D., M.S., T.H., A.Z. and T.S.; resources, T.H., M.S. and T.S.; software, T.S.; validation, M.D., M.S., T.S., A.Z. and T.H.; formal analysis, M.D. and A.Z.; investigation, M.D., M.S. and T.S.; writing—original draft preparation, M.D. and A.Z.; writing—review and editing, M.D., M.S., T.S., A.Z. and T.H.; visualization, M.D.; supervision, A.Z, T.H. and M.S. All authors have read and agreed to the published version of the manuscript.

Funding: This research was supported by the Polish National Agency for Academic Exchange, PROM Programm, International scholarship exchange of PhD candidates and academic staff.

Acknowledgments: The authors would like to thank all those whose contributed to preparing this paper, i.e., the research group from the Department of Metallic Biomaterials Institute of Biomaterials and Bioengineering, Tokyo Medical and Dental University (especially Manaka Tomoyo and Tatsuya Tsuyuzaki), and the research group from the Department of Materials Engineering and Bonding, Gdańsk University of Technology (especially Grzegorz Gajowiec and Michał Bartmański) for their technical and scientific support. Moreover, our appreciation, in particular, goes to the Sincor Company for support in thickness measurements.

Conflicts of Interest: The authors declare no conflict of interest.

References

1. Niinomi, M.; Nakai, M.; Hieda, J. Development of new metallic alloys for biomedical applications. *Acta Biomater.* **2012**, *8*, 3888–3903. [CrossRef]
2. Yu, S.; Yu, Z.T.; Wang, G.; Han, J.Y.; Ma, X.Q.; Dargusch, M.S. Preparation and osteoinduction of active micro-arc oxidation films on Ti-3Zr-2Sn-3Mo-25Nb alloy. *Trans. Nonferrous Met. Soc. China (Engl. Ed.)* **2011**, *21*, 573–580. [CrossRef]

3. Legostaeva, E.V.; Sharkeev, Y.P.; Epple, M.; Prymak, O. Structure and properties of microarc calcium phosphate coatings on the surface of titanium and zirconium alloys. *Russ. Phys. J.* **2014**, *56*, 1130–1136. [CrossRef]
4. Abdel-Hady Gepreel, M.; Niinomi, M. Biocompatibility of Ti-alloys for long-term implantation. *J. Mech. Behav. Biomed. Mater.* **2013**, *20*, 407–415. [CrossRef]
5. Elias, L.M.; Schneider, S.G.; Schneider, S.; Silva, H.M.; Malvisi, F. Microstructural and mechanical characterization of biomedical Ti–Nb–Zr(–Ta) alloys. *Mater. Sci. Eng. A* **2006**, *432*, 108–112. [CrossRef]
6. Ha, J.Y.; Tsutsumi, Y.; Doi, H.; Nomura, N.; Kim, K.H.; Hanawa, T. Enhancement of calcium phosphate formation on zirconium by micro-arc oxidation and chemical treatments. *Surf. Coat. Technol.* **2011**, *205*, 4948–4955. [CrossRef]
7. Alves, S.A.; Bayón, R.; de Viteri, V.S.; Garcia, M.P.; Igartua, A.; Fernandes, M.H.; Rocha, L.A. Tribocorrosion Behavior of Calcium- and Phosphorous-Enriched Titanium Oxide Films and Study of Osteoblast Interactions for Dental Implants. *J. Bio- Tribo-Corros.* **2015**, *1*. [CrossRef]
8. Zaraska, L.; Gawlak, K.; Gurgul, M.; Gilek, D.; Kozieł, M.; Socha, R.P.; Sulka, G.D. Morphology of nanoporous anodic films formed on tin during anodic oxidation in less commonly used acidic and alkaline electrolytes. *Surf. Coat. Technol.* **2019**, *362*, 191–199. [CrossRef]
9. Zhang, L.-C.; Chen, L.-Y.; Wang, L. Surface Modification of Titanium and Titanium Alloys: Technologies, Developments, and Future Interests. *Adv. Eng. Mater.* **2020**, *22*, 1901258. [CrossRef]
10. Tao, X.J.; Li, S.J.; Zheng, C.Y.; Fu, J.; Guo, Z.; Hao, Y.L.; Yang, R.; Guo, Z.X. Synthesis of a porous oxide layer on a multifunctional biomedical titanium by micro-arc oxidation. *Mater. Sci. Eng. C* **2009**, *29*, 1923–1934. [CrossRef]
11. Alves, A.C.; Costa, A.I.; Toptan, F.; Alves, J.L.; Leonor, I.; Ribeiro, E.; Reis, R.L.; Pinto, A.M.P.; Fernandes, J.C.S. Effect of bio-functional MAO layers on the electrochemical behaviour of highly porous Ti. *Surf. Coat. Technol.* **2020**, *386*, 125487. [CrossRef]
12. Alves, A.C.; Wenger, F.; Ponthiaux, P.; Celis, J.P.; Pinto, A.M.; Rocha, L.A.; Fernandes, J.C.S. Corrosion mechanisms in titanium oxide-based films produced by anodic treatment. *Electrochim. Acta* **2017**, *234*, 16–27. [CrossRef]
13. Wei, D.; Zhou, Y.; Jia, D.; Wang, Y. Effect of applied voltage on the structure of microarc oxidized TiO$_2$-based bioceramic films. *Mater. Chem. Phys.* **2007**, *104*, 177–182. [CrossRef]
14. Correa, D.R.N.; Rocha, L.A.; Ribeiro, A.R.; Gemini-Piperni, S.; Archanjo, B.S.; Achete, C.A.; Werckmann, J.; Afonso, C.R.M.; Shimabukuro, M.; Doi, H.; et al. Growth mechanisms of Ca- and P-rich MAO films in Ti-15Zr-xMo alloys for osseointegrative implants. *Surf. Coat. Technol.* **2018**, *344*, 373–382. [CrossRef]
15. Yerokhin, A.; Parfenov, E.V.; Matthews, A. In situ impedance spectroscopy of the plasma electrolytic oxidation process for deposition of Ca- and P-containing coatings on Ti. *Surf. Coat. Technol.* **2016**, *301*, 54–62. [CrossRef]
16. Sedelnikova, M.B.; Komarova, E.G.; Sharkeev, Y.P.; Tolkacheva, T.V.; Khlusov, I.A.; Litvinova, L.S.; Yurova, K.A.; Shupletsova, V.V. Comparative investigations of structure and properties of micro-arc wollastonite-calcium phosphate coatings on titanium and zirconium-niobium alloy. *Bioact. Mater.* **2017**, *2*, 177–184. [CrossRef]
17. Wang, Y.; Lou, J.; Zeng, L.; Xiang, J.; Zhang, S.; Wang, J.; Xiong, F.; Li, C.; Zhao, Y.; Zhang, R. Osteogenic potential of a novel microarc oxidized coating formed on Ti6Al4V alloys. *Appl. Surf. Sci.* **2017**, *412*, 29–36. [CrossRef]
18. Sowa, M.; Łastówka, D.; Kukharenko, A.I.; Korotin, D.M.; Kurmaev, E.Z.; Cholakh, S.O.; Simka, W. Characterisation of anodic oxide films on zirconium formed in sulphuric acid: XPS and corrosion resistance investigations. *J. Solid State Electrochem.* **2017**, *21*, 203–210. [CrossRef]
19. Du, Q.; Wei, D.; Wang, Y.; Cheng, S.; Liu, S.; Zhou, Y.; Jia, D. The effect of applied voltages on the structure, apatite-inducing ability and antibacterial ability of micro arc oxidation coating formed on titanium surface. *Bioact. Mater.* **2018**, *3*, 426–433. [CrossRef]
20. Wang, J.; Pan, Y.; Feng, R.; Cui, H.; Gong, B.; Zhang, L.; Gao, Z.; Cui, X.; Zhang, H.; Jia, Z. Effect of electrolyte composition on the microstructure and bio-corrosion behavior of micro-arc oxidized coatings on biomedical Ti6Al4V alloy. *J. Mater. Res. Technol.* **2020**, *9*, 1477–1490. [CrossRef]
21. Effect of Micro-Arc Oxidation Time and Applied Voltage on Formation of Strontium- and Silicon-Incorporated Biocoatings—NASA/ADS. Available online: https://ui.adsabs.harvard.edu/abs/2018JPhCS1115c2074K/abstract (accessed on 26 April 2020).

22. Han, Y.; Hong, S.H.; Xu, K. Structure and in vitro bioactivity of titania-based films by micro-arc oxidation. *Surf. Coat. Technol.* **2003**, *168*, 249–258. [CrossRef]
23. Tsutsumi, Y.; Niinomi, M.; Nakai, M.; Tsutsumi, H.; Doi, H.; Nomura, N.; Hanawa, T. Micro-arc oxidation treatment to improve the hard-tissue compatibility of Ti–29Nb–13Ta–4.6Zr alloy. *Appl. Surf. Sci.* **2012**, *262*, 34–38. [CrossRef]
24. Tsutsumi, Y.; Ashida, M.; Nakahara, K.; Serizawa, A.; Doi, H.; Grandini, C.R.; Rocha, L.A.; Hanawa, T. Micro arc oxidation of Ti–15Zr–7.5Mo alloy. *Mater. Trans.* **2016**, *57*, 2015–2019. [CrossRef]
25. Wang, C.; Ma, F.; Liu, P.; Chen, J.; Liu, X.; Zhang, K.; Li, W.; Han, Q. The influence of alloy elements in Ti–6Al–4V and Ti–35Nb–2Ta–3Zr on the structure, morphology and properties of MAO coatings. *Vacuum* **2018**, *157*, 229–236. [CrossRef]
26. Liu, S.; Li, B.; Liang, C.; Wang, H.; Qiao, Z. Formation mechanism and adhesive strength of a hydroxyapatite/TiO$_2$ composite coating on a titanium surface prepared by micro-arc oxidation. *Appl. Surf. Sci.* **2016**, *362*, 109–114. [CrossRef]
27. Dziaduszewska, M.; Wekwejt, M.; Bartmański, M.; Pałubicka, A.; Gajowiec, G.; Seramak, T.; Osyczka, A.M.; Zieliński, A. The Effect of Surface Modification of Ti13Zr13Nb Alloy on Adhesion of Antibiotic and Nanosilver-Loaded Bone Cement Coatings Dedicated for Application as Spacers. *Materials* **2019**, *12*, 2964. [CrossRef]
28. Yao, J.; Wang, Y.; Wu, G.; Sun, M.; Wang, M.; Zhang, Q. Growth characteristics and properties of micro-arc oxidation coating on SLM-produced TC4 alloy for biomedical applications. *Appl. Surf. Sci.* **2019**, *479*, 727–737. [CrossRef]
29. Dziaduszewska, M.; Zielinski, A. Titanium Scaffolds-Hopes and Limitations. *Am. J. Biomed. Sci. Res.* **2019**, *4*, 390–391. [CrossRef]
30. Zhang, B.; Pei, X.; Zhou, C.; Fan, Y.; Jiang, Q.; Ronca, A.; D'Amora, U.; Chen, Y.; Li, H.; Sun, Y.; et al. The biomimetic design and 3D printing of customized mechanical properties porous Ti6Al4V scaffold for load-bearing bone reconstruction. *Mater. Des.* **2018**, *152*, 30–39. [CrossRef]
31. Yang, J.; Yu, H.; Yin, J.; Gao, M.; Wang, Z.; Zeng, X. Formation and control of martensite in Ti-6Al-4V alloy produced by selective laser melting. *Mater. Des.* **2016**, *108*, 308–318. [CrossRef]
32. van Hengel, I.A.J.; Riool, M.; Fratila-Apachitei, L.E.; Witte-Bouma, J.; Farrell, E.; Zadpoor, A.A.; Zaat, S.A.J.; Apachitei, I. Selective laser melting porous metallic implants with immobilized silver nanoparticles kill and prevent biofilm formation by methicillin-resistant Staphylococcus aureus. *Biomaterials* **2017**, *140*, 1–15. [CrossRef]
33. Liu, Z.; Jia, Z.; Lv, J.; Yin, C.; Cai, H.; Song, C.; Leng, H.; Zheng, Y.; Liu, Z.; Cheng, Y. Hierarchical Micropore/Nanorod Apatite Hybrids In-Situ Grown from 3-D Printed Macroporous Ti6Al4V Implants with Improved Bioactivity and Osseointegration. *J. Mater. Sci. Technol.* **2017**, *33*, 179–186. [CrossRef]
34. Nyan, M.; Tsutsumi, Y.; Oya, K.; Doi, H.; Nomura, N.; Kasugai, S.; Hanawa, T. Synthesis of novel oxide layers on titanium by combination of sputter deposition and micro-arc oxidation techniques. *Dent. Mater. J.* **2011**, *30*, 754–761. [CrossRef]
35. Shimabukuro, M.; Tsutsumi, Y.; Nozaki, K.; Chen, P.; Yamada, R.; Ashida, M.; Doi, H.; Nagai, A.; Hanawa, T. Chemical and biological roles of zinc in a porous titanium dioxide layer formed by micro-arc oxidation. *Coatings* **2019**, *9*, 705. [CrossRef]
36. Tanaka, Y.; Kobayashi, E.; Hiromoto, S.; Asami, K.; Imai, H.; Hanawa, T. Calcium phosphate formation on titanium by low-voltage electrolytic treatments. *J. Mater. Sci. Mater. Med.* **2007**, *18*, 797–806. [CrossRef]
37. Oliver, W.C.; Pharr, G.M. An improved technique for determining hardness and elastic modulus using load and displacement sensing indentation experiments. *J. Mater. Res.* **1992**, *7*, 1564–1583. [CrossRef]
38. Ehlert, M.; Radtke, A.; Jedrzejewski, T.; Roszek, K.; Bartmanski, M.; Piszczek, P. In vitro studies on nanoporous, nanotubular and nanosponge-like titania coatings, with the use of adipose-derived stem cells. *Materials* **2020**, *13*, 1574. [CrossRef]
39. Li, J.; Cai, H.; Jiang, B. Growth mechanism of black ceramic layers formed by microarc oxidation. *Surf. Coat. Technol.* **2007**, *201*, 8702–8708. [CrossRef]
40. Qian, B.-Y.; Miao, W.; Qiu, M.; Gao, F.; Hu, D.-H.; Sun, J.-F.; Wu, R.-Z.; Krit, B.; Betsofen, S. Influence of Voltage on the Corrosion and Wear Resistance of Micro-Arc Oxidation Coating on Mg–8Li–2Ca Alloy. *Acta Metall. Sin. (English Lett.)* **2019**, *32*, 194–204. [CrossRef]

41. Sobolev, A.; Kossenko, A.; Borodianskiy, K. Study of the effect of current pulse frequency on Ti-6Al-4V alloy coating formation by micro arc oxidation. *Materials* **2019**, *12*, 3983. [CrossRef]
42. Wang, J.H.; Wang, J.; Lu, Y.; Du, M.H.; Han, F.Z. Effects of single pulse energy on the properties of ceramic coating prepared by micro-arc oxidation on Ti alloy. *Appl. Surf. Sci.* **2015**, *324*, 405–413. [CrossRef]
43. Heimann, R.B. Structure, properties, and biomedical performance of osteoconductive bioceramic coatings. *Surf. Coat. Technol.* **2013**, *233*, 27–38. [CrossRef]
44. Kashyap, S.; Griep, K.; Nychka, J.A. Crystallization kinetics, mineralization and crack propagation in partially crystallized bioactive glass 45S5. *Mater. Sci. Eng. C* **2011**, *31*, 762–769. [CrossRef]
45. Ishizawa, H.; Ogino, M. Formation and characterization of anodic titanium oxide films containing Ca and P. *J. Biomed. Mater. Res.* **1995**, *29*, 65–72. [CrossRef]
46. Chen, L.; Jin, X.; Qu, Y.; Wei, K.; Zhang, Y.; Liao, B.; Xue, W. High temperature tribological behavior of microarc oxidation film on Ti-39Nb-6Zr alloy. *Surf. Coat. Technol.* **2018**, *347*, 29–37. [CrossRef]
47. Karbowniczek, J.; Muhaffel, F.; Cempura, G.; Cimenoglu, H.; Czyrska-Filemonowicz, A. Influence of electrolyte composition on microstructure, adhesion and bioactivity of micro-arc oxidation coatings produced on biomedical Ti6Al7Nb alloy. *Surf. Coat. Technol.* **2017**, *321*, 97–107. [CrossRef]
48. Kazek-Kesik, A.; Krok-Borkowicz, M.; Jakóbik-Kolon, A.; Pamuła, E.; Simka, W. Biofunctionalization of Ti-13Nb-13Zr alloy surface by plasma electrolytic oxidation. Part II. *Surf. Coat. Technol.* **2015**, *276*, 23–30. [CrossRef]
49. Wang, Y.H.; Ouyang, J.H.; Liu, Z.G.; Wang, Y.M.; Wang, Y.J. Microstructure and high temperature properties of two-step voltage-controlled MAO ceramic coatings formed on Ti 2 AlNb alloy. *Appl. Surf. Sci.* **2014**, *307*, 62–68. [CrossRef]
50. Chien, C.S.; Hung, Y.C.; Hong, T.F.; Wu, C.C.; Kuo, T.Y.; Lee, T.M.; Liao, T.Y.; Lin, H.C.; Chuang, C.H. Preparation and characterization of porous bioceramic layers on pure titanium surfaces obtained by micro-arc oxidation process. *Appl. Phys. A Mater. Sci. Process.* **2017**, *123*, 1–10. [CrossRef]
51. Wang, Y.M.; Guo, J.W.; Zhuang, J.P.; Jing, Y.B.; Shao, Z.K.; Jin, M.S.; Zhang, J.; Wei, D.Q.; Zhou, Y. Development and characterization of MAO bioactive ceramic coating grown on micro-patterned Ti6Al4V alloy surface. *Appl. Surf. Sci.* **2014**, *299*, 58–65. [CrossRef]
52. Planell, J.A.; Navarro, M.; Altankov, G.; Aparicio, C.; Engel, E.; Gil, J.; Ginebra, M.P.; Lacroix, D. Materials Surface Effects on Biological Interactions. In *Advances in Regenerative Medicine: Role of Nanotechnology, and Engineering Principles*; Springer: Dordrecht, The Netherlands, 2010; pp. 233–252. [CrossRef]
53. Love, G.; Scott, V.D.; Dennis, N.M.T.; Laurenson, L. Sources of contamination in electron optical equipment. *Scanning* **1981**, *4*, 32–39. [CrossRef]
54. Tsai, M.T.; Chang, Y.Y.; Huang, H.L.; Wu, Y.H.; Shieh, T.M. Micro-arc oxidation treatment enhanced the biological performance of human osteosarcoma cell line and human skin fibroblasts cultured on titanium–zirconium films. *Surf. Coat. Technol.* **2016**, *303*, 268–276. [CrossRef]
55. Cordero-Arias, L.; Cabanas-Polo, S.; Gao, H.; Gilabert, J.; Sanchez, E.; Roether, J.A.; Schubert, D.W.; Virtanen, S.; Boccaccini, A.R. Electrophoretic deposition of nanostructured-TiO_2/chitosan composite coatings on stainless steel. *RSC Adv.* **2013**, *3*, 11247. [CrossRef]
56. Durdu, S. Characterization, bioactivity and antibacterial properties of copper-based TiO_2 bioceramic coatings fabricated on titanium. *Coatings* **2019**, *9*, 1. [CrossRef]
57. Wang, L.; Wang, K.; Erkan, N.; Yuan, Y.; Chen, J.; Nie, B.; Li, F.; Okamoto, K. Metal material surface wettability increase induced by electron beam irradiation. *Appl. Surf. Sci.* **2020**, *511*, 145555. [CrossRef]
58. Acharya, S.; Panicker, A.G.; Laxmi, D.V.; Suwas, S.; Chatterjee, K. Study of the influence of Zr on the mechanical properties and functional response of Ti-Nb-Ta-Zr-O alloy for orthopedic applications. *Mater. Des.* **2019**, *164*, 107555. [CrossRef]
59. Hanaor, D.A.H.; Sorrell, C.C. Review of the anatase to rutile phase transformation. *J. Mater. Sci.* **2011**, *46*, 855–874. [CrossRef]
60. Zhang, P.; Zhang, Z.; Li, W.; Zhu, M. Effect of Ti-OH groups on microstructure and bioactivity of TiO_2 coating prepared by micro-arc oxidation. *Appl. Surf. Sci.* **2013**, *268*, 381–386. [CrossRef]
61. Tang, H.; Tao, W.; Wang, C.; Yu, H. Fabrication of hydroxyapatite coatings on AZ31 Mg alloy by micro-arc oxidation coupled with sol-gel treatment. *RSC Adv.* **2018**, *8*, 12368–12375. [CrossRef]
62. Cheng, S.; Wei, D.; Zhou, Y.; Guo, H. Characterization and properties of microarc oxidized coatings containing Si, Ca and Na on titanium. *Ceram. Int.* **2011**, *37*, 1761–1768. [CrossRef]

63. Szesz, E.M.; de Souza, G.B.; de Lima, G.G.; da Silva, B.A.; Kuromoto, N.K.; Lepienski, C.M. Improved tribo-mechanical behavior of CaP-containing TiO$_2$ layers produced on titanium by shot blasting and micro-arc oxidation. *J. Mater. Sci. Mater. Med.* **2014**, *25*, 2265–2275. [CrossRef]
64. Bhushan, B.; Gupta, B.K.; Azarian, M.H. Nanoindentation, microscratch, friction and wear studies of coatings for contact recording applications. *Wear* **1995**, *181–183*, 743–758. [CrossRef]
65. Chen, Z. Nanoindentation of Macro-porous Materials for Elastic Modulus and Hardness Determination. In *Applied Nanoindentation in Advanced Materials*; Tiwari, A., Natarajan, S., Eds.; John Wiley & Sons, Ltd.: Hoboken, NJ, USA, 2017; pp. 135–156. [CrossRef]
66. Savchenko, N.; Sevostyanova, I.; Sablina, T.; Gömze, L.; Kulkov, S. The Influence of porosity on the Elasticity and Strength of Alumina and Zirconia Ceramics. In Proceedings of the International Conference on Physical Mecomechanics of Multilevel System, Tomsk, Russia, 3–5 September 2014; Panin, V.E., Psakhie, S.G., Fomin, V.M., Eds.; AIP Publishing LLC: Melville, NY, USA, 2014; pp. 547–550. [CrossRef]
67. Chen, Z.; Wang, X.; Bhakhri, V.; Giuliani, F.; Atkinson, A. Nanoindentation of Porous Bulk and Thin Films of La$_{0.6}$Sr$_{0.4}$Co$_{0.2}$Fe$_{0.8}$O$_{3-\delta}$. *Acta Mater.* **2013**, *61*, 5720–5734. [CrossRef]
68. Herrmann, M.; Richter, F.; Schulz, S.E. Study of nano-mechanical properties for thin porous films through instrumented indentation: SiO$_2$ low dielectric constant films as an example. *Microelectron. Eng.* **2008**, *85*, 2172–2174. [CrossRef]
69. de Souza, G.B.; Lepienski, C.M.; Foerster, C.E.; Kuromoto, N.K.; Soares, P.; de Araújo Ponte, H. Nanomechanical and nanotribological properties of bioactive titanium surfaces prepared by alkali treatment. *J. Mech. Behav. Biomed. Mater.* **2011**, *4*, 756–765. [CrossRef]
70. Solis, J.; Zhao, H.; Wang, C.; Verduzco, J.A.; Bueno, A.S.; Neville, A. Tribological performance of an H-DLC coating prepared by PECVD. *Appl. Surf. Sci.* **2016**, *383*, 222–232. [CrossRef]
71. Zimowski, S. Wpływ twardości i modułu sprężystości powłok kompozytowych na ich odporność na zużycie. *Tribologia* **2014**, *6*, 149–160.
72. Lępicka, M.; Grldzka-Dahlke, M. The initial evaluation of performance of hard anti-wear coatings deposited on metallic substrates: Thickness, mechanical properties and adhesion measurements—A brief review. *Rev. Adv. Mater. Sci.* **2019**, *58*, 50–65. [CrossRef]
73. Huang, X.; Etsion, I.; Shao, T. Effects of elastic modulus mismatch between coating and substrate on the friction and wear properties of TiN and TiAlN coating systems. *Wear* **2015**, *338–339*, 54–61. [CrossRef]
74. Bartmański, M.; Pawłowski, Ł.; Strugała, G.; Mielewczyk-Gryń, A.; Zieliński, A. Properties of nanohydroxyapatite coatings doped with nanocopper, obtained by electrophoretic deposition on Ti13Zr13Nb alloy. *Materials* **2019**, *12*, 3741. [CrossRef]
75. Jiang, W.G.; Su, J.J.; Feng, X.Q. Effect of surface roughness on nanoindentation test of thin films. *Eng. Fract. Mech.* **2008**, *75*, 4965–4972. [CrossRef]
76. Chen, X.; Vlassak, J.J. Numerical study on the measurement of thin film mechanical properties by means of nanoindentation. *J. Mater. Res.* **2001**, *16*, 2974–2982. [CrossRef]
77. Saha, R.; Nix, W.D. Effects of the substrate on the determination of thin film mechanical properties by nanoindentation. *Acta Mater.* **2002**, *50*, 23–38. [CrossRef]
78. Xiang, Y.; Chen, X.; Tsui, T.Y.; Jang, J.I.; Vlassak, J.J. Mechanical properties of porous and fully dense low-κ dielectric thin films measured by means of nanoindentation and the plane-strain bulge test technique. *J. Mater. Res.* **2006**, *21*, 386–395. [CrossRef]
79. Chen, X.; Xiang, Y.; Vlassak, J.J. Novel technique for measuring the mechanical properties of porous materials by nanoindentation. *J. Mat. Res.* **2006**, *21*, 715–724. [CrossRef]
80. Bouzakis, K.D.; Michailidis, N.; Hadjiyiannis, S.; Skordaris, G.; Erkens, G. The effect of specimen roughness and indenter tip geometry on the determination accuracy of thin hard coatings stress-strain laws by nanoindentation. *Mater. Charact.* **2002**, *49*, 149–156. [CrossRef]
81. Aydin, I.; Bahçepinar, A.I.; Kirman, M.; Çipiloğlu, M.A. HA coating on Ti6Al7Nb alloy using an electrophoretic deposition method and surface properties examination of the resulting coatings. *Coatings* **2019**, *9*, 402. [CrossRef]
82. Cai, X.; Xu, Y.; Zhao, N.; Zhong, L.; Zhao, Z.; Wang, J. Investigation of the adhesion strength and deformation behaviour of in situ fabricated NbC coatings by scratch testing. *Surf. Coat. Technol.* **2016**, *299*, 135–142. [CrossRef]

83. Krawiec, H.; Vignal, V.; Krystianiak, A.; Gaillard, Y.; Zimowski, S. Mechanical properties and corrosion behaviour after scratch and tribological tests of electrodeposited Co-Mo/TiO$_2$ nano-composite coatings. *Appl. Surf. Sci.* **2019**, *475*, 162–174. [CrossRef]
84. Lenz, B.; Hasselbruch, H.; Großmann, H.; Mehner, A. Application of CNN networks for an automatic determination of critical loads in scratch tests on a-C:H:W coatings. *Surf. Coat. Technol.* **2020**, *393*, 125764. [CrossRef]
85. Randall, N.X. The current state-of-the-art in scratch testing of coated systems. *Surf. Coat. Technol.* **2019**, *380*, 125092.
86. Su, Y.; Komasa, S.; Sekino, T.; Nishizaki, H.; Okazaki, J. Nanostructured Ti6Al4V alloy fabricated using modified alkali-heat treatment: Characterization and cell adhesion. *Mater. Sci. Eng. C* **2016**, *59*, 617–623. [CrossRef]
87. Zhu, X.; Kim, K.H.; Jeong, Y. Anodic oxide films containing Ca and P of titanium biomaterial. *Biomaterials* **2001**, *22*, 2199–2206. [CrossRef]

© 2020 by the authors. Licensee MDPI, Basel, Switzerland. This article is an open access article distributed under the terms and conditions of the Creative Commons Attribution (CC BY) license (http://creativecommons.org/licenses/by/4.0/).

Article

Comparison of Properties of the Hybrid and Bilayer MWCNTs—Hydroxyapatite Coatings on Ti Alloy

Beata Majkowska-Marzec [1], Dorota Rogala-Wielgus [1], Michał Bartmański [1], Bartosz Bartosewicz [2] and Andrzej Zieliński [1],*

1. Department of Materials Engineering and Bonding, Gdańsk University of Technology, Narutowicza 11/12 str., 80-233 Gdańsk, Poland; beamajko@pg.edu.pl (B.M.-M.); dorota.wielgus@pg.edu.pl (D.R.-W.); michal.bartmanski@pg.edu.pl (M.B.)
2. Institute of Optoelectronics, Military University of Technology, gen. Sylwestra Kaliskiego 2 str., 00-908 Warsaw, Poland; bartosz.bartosewicz@wat.edu.pl
* Correspondence: andrzej.zielinski@pg.edu.pl; Tel.: +48-501-329-368; Fax: +48-583471815

Received: 16 September 2019; Accepted: 30 September 2019; Published: 4 October 2019

Abstract: Carbon nanotubes are proposed for reinforcement of the hydroxyapatite coatings to improve their adhesion, resistance to mechanical loads, biocompatibility, bioactivity, corrosion resistance, and antibacterial protection. So far, research has shown that all these properties are highly susceptible to the composition and microstructure of coatings. The present research is aimed at studies of multi-wall carbon nanotubes in three different combinations: multi-wall carbon nanotubes layer, bilayer coating composed of multi-wall carbon nanotubes deposited on nanohydroxyapatite deposit, and hybrid coating comprised of simultaneously deposited nanohydroxyapatite, multi-wall carbon nanotubes, nanosilver, and nanocopper. The electrophoretic deposition method was applied for the fabrication of the coatings. Atomic force microscopy, scanning electron microscopy and X-ray electron diffraction spectroscopy, and measurements of water contact angle were applied to study the chemical and phase composition, roughness, adhesion strength and wettability of the coatings. The results show that the pure multi-wall carbon nanotubes layer possesses the best adhesion strength, mechanical properties, and biocompatibility. Such behavior may be attributed to the applied deposition method, resulting in the high hardness of the coating and high adhesion of carbon nanotubes to the substrate. On the other hand, bilayer coating, and hybrid coating demonstrated insufficient properties, which could be the reason for the presence of soft porous hydroxyapatite and some agglomerates of nanometals in prepared coatings.

Keywords: hardness; adhesion; hydroxyapatite; carbon nanotubes; titanium; biomedical applications

1. Introduction

Carbon nanotube coatings (CNTs) demonstrate unique mechanical and biological properties. Thanks to this, they are increasingly applied in medicine and diagnostics, including tissue engineering [1,2]. These two-dimensional carbon structures are used, among others, to functionalize materials designed for implants, where CNTs can support osseointegration [3,4].

The biocompatibility of CNTs in orthopedic applications was established by in vitro studies, which showed accelerated bone growth and increased proliferation and differentiation of osteoblasts [5–10]. The most popular kind of metal substrate for CNTs is titanium [3,11–18], which combines some beneficial mechanical properties and biocompatibility with a chemical in vivo susceptibility [11]. Several studies evaluated the body's reaction in the presence of carbon nanotubes, demonstrating a high vitality of osteoblasts compared to the pure titanium substrate [14–16,19]. The ceramic coating consisting of multiple functionalized CNTs with carboxyl groups and hydroxyapatite (HAp) was reported to enhance mechanical properties and biological adhesion, as well as the response of osteoblasts [20–23]. CNTs have a unique chemical structure, so it could serve as a carrier, e.g., of an antibiotic [11] or other substances [21], able to prevent or cure potential infection in place of implantation and protect against implant rejection.

Nevertheless, possible toxicity of nanomaterials limits CNTs' medical applications [24], even though reports to date are scarce and inconclusive, e.g., for the neurotoxic effects of CNTs [25], or CNTs formation of reactive oxygen species [26]. There are more reports on the lack of adverse effects of CNTs than on their long-term toxicity [24]. Recent toxicological studies performed with the liver and kidney cells showed no adverse outcome [23]. In another study, neither SWCNTs (Single-Wall CNTs) nor MWCNTs (Mulit-Wall CNTs) demonstrated in vitro cytotoxicity for fibroblasts and hippocampal cells [27]. Moreover, CNTs are extensively investigated as components of biocoatings [2], e.g., for the Mg-phosphate coating reinforced by SiC nanowire-CNTs [28], HAp-CNTs composite coating on Mg alloy [29], SWCNTs/HAp and MWCNTs on Ti and its alloys [14–17]. Thus, the cytotoxicity of CNTs is generally assumed to be negligible. Even the use of both SWCNTs and MWCNTs as a base liquid with human blood was reported [30].

Pure SWCNTs, as well as MWCNTs coatings, have been obtained [31–33], but the composite or hybrid layers are even more extensively developed as, e.g., CNTs–HAp [13,22,23,34–37] or CNTs–graphene oxide [38]. Most frequently, the electrophoretic deposition (EPD) [12,14,39–41], electrocathodic deposition [15,16,20] or chemical vapor deposition (CVD) processes [42] are applied to prepare CNTs coating.

In the case of materials intended for dental or orthopedic load-bearing implants, adhesion and bond strength are essential, especially at the stage of implantation and ingrowth of human tissue. As Gopi et al. observed, the addition of low amounts of MWCNTs (0.5 and 1 mass pct.) increases hardness, and Young's modulus of the sol-gel derived HAp/MWCNTs coatings [15]. Nevertheless, the concentration of 1% and 2% of CNTs in the composite did not affect adhesion strength significantly and reached 24.2 and 22.4 MPa, respectively [16]. Mukherjee et al. reported an improvement of fracture toughness, flexural and impact strength values for MWCNTs-reinforced HAp [23]. The interfacial shear strength and the maximum load-bearing capacity of the tested CNTs–Ti interfaces were assessed at 37.8 MPa and 245 nN, respectively [43]. The tape tests displayed high adhesion strength (class 5B) for CNTs–(Zn)HAp coating [18]. In another work, the addition of CNTs to the Ti–HAp composite improved both the adhesion strength and hardness for the Ni–Ti substrate [35]. The highest bonding strength of 25.7 MPa was reported for the SWNTs/HAp coating and was nearly 70% more elevated than that of the pure HAp coating [17]. Better mechanical properties were observed for new complex hybrid materials, such as MWCNTs-HAp [20] and fluorohydroxyapatite (FA)-CNTs coating deposited on the TiO_2 nanotubular layer [44]. In the last case, CNTs served as a reinforcement, because of their higher elastic modulus compared to the FA matrix. The homogeneous distribution of decorated CNTs resulted in the robust interface between FA and CNTs. CNT-reinforced HAp composites had substantially better bending strength and fracture toughness than pure HAp [45]. Thus, the presence of MWCNTs evidently improves cohesion, but its adhesion to titanium substrate is less known and may be dependent on the specific architecture of a coating.

In sum, all the investigated material features such as biocompatibility, adhesion strength, and corrosion resistance are highly sensitive to coating composition and microstructure. The objective of the present research is to assess the adhesion strength, mechanical properties of coatings and wettability for the pure MWCNTs layer, hybrid CNTs–nanohydroxyapatite (nanoHAp) coating, and composite coating. Three different methods of deposition of carbon nanotubes were applied to obtain such coatings: (i) deposition of MWCNTs on a substrate surface, (ii) deposition of nanoHAp coating followed by the deposition of MWCNTs, and (iii) joint deposition of a mixture of nanoHAp, MWCNTs, and some nanometals. The purpose of such a choice lies in the fundamental importance of adhesion for the application of such surface treatment of load-bearing implants, because of the high stresses imposed on them during implantation surgery and the post-implantation period.

2. Materials and Methods

2.1. Preparation of Substrate Surfaces

The Ti13Nb13Zr alloy of the composition shown in Table 1 was used as a substrate. Specimens with a 40 mm diameter were cut from the rods. The surface was ground using abrasive paper SiC up to grit # 800. Then, the samples were rinsed with acetone, distilled water, air-dried, pickled in 5% HF for 30 s to remove oxide layers from the surface and finally rinsed with distilled water.

Table 1. Chemical composition of the Ti13Nb13Zr alloy.

Element	Nb	Zr	Fe	C	H	O	S	Hf	Ti
wt. pct.	13.18	13.49	0.085	0.035	0.004	0.078	<0.001	0.055	rem.

2.2. Preparation of CNTs' Suspension

To prepare the coatings, MWCNTs (3D-nano, number of walls 3–15, outer diameter 5–20 nm, inner diameter 2–6 nm, and length 1–10 μm) were functionalized in a mixture of concentrated sulfuric and nitric acid to add carboxyl groups and to provide a negative charge on the surface of carbon nanotubes. Four hundred and eighty grams of powder was annealed in a vacuum furnace (PROTHERM PC442, Ankara, Turkey) for 8 h at 400 °C and then dispersed in deionized water in an ultrasonic homogenizer (Bandelin Sonopuls HD 2070, Berlin, Germany). The suspension was added to 200 mL of mixed H_2SO_4 and at a ratio of 3:1 v/v and heated at 70 °C for 2 h [3]. To prepare the suspension of carbon nanotubes, the reaction mixture was centrifuged and washed several times with water or isopropanol until a neutral pH was reached. The concentration of CNTs in the obtained suspension was 0.27 wt.% in water and 0.4 wt.% in isopropanol. Final suspensions were sonicated for 1 min using an ultrasonic homogenizer to disperse the CNTs well after centrifugation. To prepare the mixed coating (m0.4CNT), 1.25 mL of 0.4 wt.% suspension of carbon nanotubes in isopropanol and 0.1 g of nanohydroxyapatite (grain size distribution approximately 20 nm, 99.8% purity, MKnano, Missisauga, Canada) were dispersed in 100 mL of ethyl alcohol (99.8% purity, Sigma Aldrich, St. Louis, MI, USA) and then mixed with 0.005 g of nanosilver (grain size distribution approx. 30 nm, Hongwu International Group Ltd., Guangzhou, China) and 0.005 g of nanocopper (grain size distribution approximately 80 nm, Hongwu International Group Ltd.) before carrying out the electrophoretic deposition (EPD) process.

2.3. Deposition of Coatings

The electrophoretic deposition (EPD) method was used to prepare the coatings. Their synthesis parameters are shown in Table 2. The Ti13Nb13Zr substrate was used as an anode and platinum as a counter electrode. The electrodes were placed parallel to each other at a distance of 5 mm and connected to a DC power source (MCP/SPN110-01C, Shanghai MCP Corp., Shanghai, China). The coatings were heated in a tubular furnace (PROTHERM PC442) from room temperature to 800 °C at a rate of 200 °C/h and cooled to room temperature with the oven.

Table 2. Parameters of synthesis of the coatings with multi-walled carbon nanotubes.

Coating	Synthesis Stage	Concentration of MWCNTs [%]	Duration of EPD [min]	EDP Voltage [V]	Temperature [°C]
0.27CNT	EPD of MWCNTs	0.27	2	11	ambient
H0.27CNT	EPD of nanoHAp	–	2	30	ambient
	Sintering	–	120	–	800
	EPD of MWCNTs	0.27	2	30	ambient
m0.4CNT	EPD of MWCNTs, nanoHAp, nanosilver, nanocopper	0.4	2	30	ambient
	Sintering	–	120	–	800

2.4. Structure and Morphology

An atomic force microscope (AFM NaniteAFM, Nanosurf, Bracknell, Great Britain) was used to study the surface topography. The examinations were performed in the non-contact mode at 20 mN force. The average roughness index S_a values were estimated based on 512 lines made in the area of 80.4×80.4 μm^2.

The specimens' surfaces were observed using a high-resolution scanning electron microscope (SEM JEOL JSM-7800F, Tokyo, Japan) with a LED detector, at a 5 kV acceleration voltage.

The chemical composition of the coatings was investigated by an X-ray energy dispersive spectrometer (EDS Edax Inc., Mahwah, NJ, USA).

2.5. Nanomechanical Studies

Nanoindentation tests were performed with the NanoTest™ Vantage (Micro Materials, Wrexham, Great Britain) using a Berkovich three-sided pyramidal diamond. Twenty-five (5 × 5) measurements were carried out on each sample. The maximum applied force was 10 mN, the loading and unloading times were set up at 20 s and the dwell period at maximum load was 10 s. The distance between the subsequent indents was 20 μm. During the indent, the load–displacement curve was determined using the Oliver and Pharr method. Based on the load–penetration depth curves, the surface hardness (H) and Young's modulus (E) were calculated using integrated software. Estimating Young's modulus (E), the Poisson's ratio of 0.25 was assumed for carbon nanotube coatings and 0.36 for Ti13Nb13Zr.

Nanoscratch tests were performed with NanoTest™ Vantage (Micro Materials) using a Berkovich three-sided pyramidal diamond. The scratch tests were made by increasing the load from 0 to 200 mN at a loading rate of 1.3 mN/s at a distance of 500 μm. The adhesion of the coating was assessed based on the observation of an abrupt change in frictional force during the test.

2.6. Contact Angle Studies

Water contact angle measurements were carried out by falling drop method using a contact angle instrument (Contact Angle Goniometer, Zeiss, Oberkochen, Germany) at room temperature 10 s after drop out.

3. Results and Discussion

3.1. Structure and Morphology

Figure 1A,B show the AFM images of the topography of the native material, where substantial roughness of the surface after grinding and etching may be observed. After etching, the native material shows higher surface roughness.

Figure 1C–E illustrate the surface topography of the examined coatings composed of, respectively, carbon nanotubes (0.27CNT), carbon nanotubes deposited on the nanohydroxyapatite layer (H0.27CNT) and the composite of carbon nanotubes, nanohydroxyapatite, nanosilver and nanocopper (m0.4CNT). The addition of nanoHAp significantly increases roughness. Such a result may mean that the HAp penetrates through free spaces among nanotubes only a small amount. Fathyunes et al. (2018) proposed that during electrodeposition the water reduction produces hydroxyl ions, causing an increase in the pH near the cathode. Then, the CaP ceramics become insoluble and precipitate on the surface of the titanium substrate [46].

Figure 2 presents the SEM images of the surface topography of carbon nanotubes (0.27CNT), hybrid coating (H0.27CNT), and composite coating (m0.4CNT) together with their EDS spectra, confirming the formation of individual coatings. The carbon nanotubes can be distinguished for each specimen.

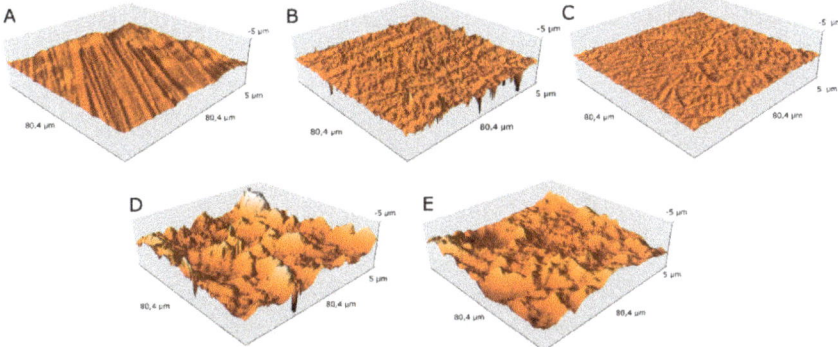

Figure 1. Atomic force microscope (AFM) surface topography of the: (**A**) reference sample—native material after grinding (MR), (**B**) reference sample—after etching (MRe), (**C**) sample of carbon nanotubes (0.27CNT), (**D**) sample of carbon nanotubes deposited on the nanohydroxyapatite layer(H0.27CNT), (**E**) sample of composite of carbon nanotubes, nanohydroxyapatite, nanosilver and nanocopper (m0.4CNT).

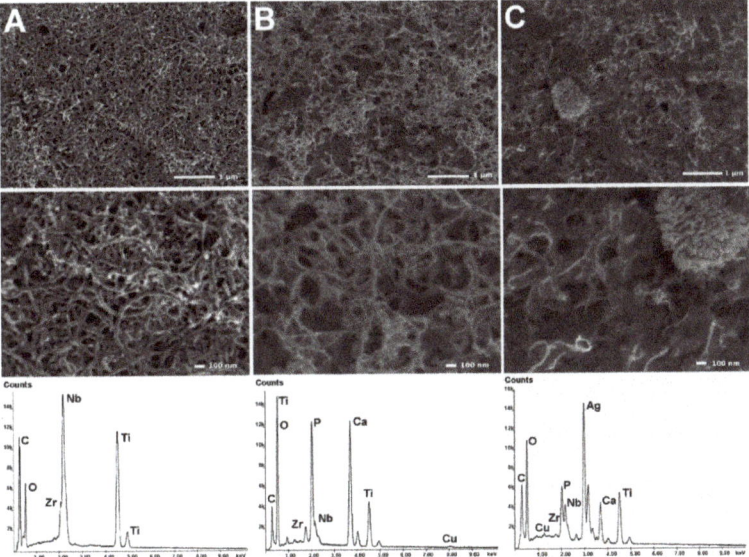

Figure 2. SEM surface topography with the energy dispersive spectrometer (EDS) spectrum of the sample: (**A**) 0.27CNT, (**B**) H0.27CNT, (**C**) m0.4CNT.

The SEM images (Figure 2) demonstrate a more uniform distribution of carbon nanotubes for the 0.27CNT sample than for the H0.27CNT sample. On the surface of the m0.4CNT sample, many agglomerates are present, as a result of the simultaneous deposition of nanosilver and nanocopper, which do not move into a bulk, but are absorbed on the HAp coating, resulting in decreased roughness (Table 3). The roughness values are similar to those previously reported [14].

Table 3. Surface roughness of the native material and deposited coatings.

Sample	Roughness S_a [µm]
MR	0.203
MRe	0.256
0.27CNT	0.098
H0.27CNT	0.980
M0.4CNT	0.618

3.2. Mechanical Studies

3.2.1. Nanoindentation

Figure 3 shows the load–displacement hysteresis curve as an example of the results of nanoindentation tests. Three stages of the nanoindentation test are observed: the raising load to a maximum value, the pause (to stabilize the probe at maximum depth), and offloading. An irregularity in the form of a step due to temperature drift, adjusted at the end of nanoindentation, can be observed.

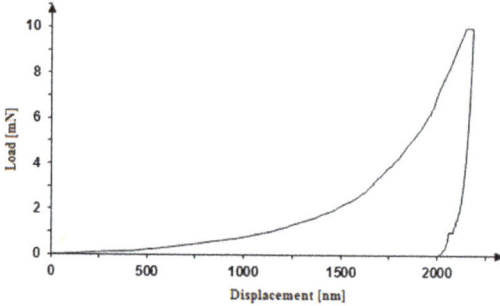

Figure 3. Nanoindentation load–displacement curve obtained for pure CNT coating (0.27CNT).

The exact values of measured mechanical properties are listed in Table 4. The lowest hardness values are demonstrated for H0.26CNT and m0.4CNT coatings due to the significant effect of etching on hardness. A 53-fold difference in the hardness is noticed between polished and etched specimens. The CNTs alone have the best hardness and wear resistance. The addition of softer nanoHAp decreases mechanical strength. The composite coating displays improved behavior, presumably because of a different coating architecture and the decisive role of metal nanoparticles. The hardness here measured is lower than the values previously reported in [44] as 0.37 to 0.58 GPa.

The elastic modulus values observed here are in line with some previous results for titanium, and much smaller for CNTs coating, so far reported as 60 MPa [14] and 113–130 MPa [16]. An increase in Young's modulus from 15 to 40 GPa [35] and from 12 to 19 GPa [44] was also observed. The reason for such discrepancies may be the high dependence of nanomechanical properties of the coating architecture, test parameters, and fractions of components.

Table 4. Mechanical properties and maximum indent depth for the substrate and achieved coatings.

Sample	Nanohardness [GPa]	Reduced Young's Modulus [GPa]	Young's Modulus [GPa]	Maximum Indent Depth [nm]
I	3.758 ± 1.045	116.91 ± 16.32	83.32 ± 11.63	330.92 ± 36.61
MRe	0.071 ± 0.010	14.57 ± 2.21	9.44 ± 1.43	2358.11 ± 170.08
0.27CNT	0.101 ± 0.049	18.59 ± 5.66	14.17 ± 4.32	2069.67 ± 352.57
H0.27CNT	0.022 ± 0.015	7.46 ± 3.65	5.63 ± 2.76	4264.18 ± 1150.11
m0.4CNT	0.035 ± 0.019	11.72 ± 4.31	8.88 ± 3.26	3210.02 ± 817.53

Figures 4 and 5 show the 3D distribution of Young's modulus and nanohardness for the examined samples. Compared to the other materials, the native material after grinding (MR) reveals the biggest nanohardness and Young modulus.

The 3D Young's modulus (Figure 4) and nanohardness (Figure 5) distribution graphs show a highly non-uniform rough surface. The roughest is only a polished surface, lower roughness is observed for CNTs coating, and the etched surface and two other coatings show differences in Young's modulus in the range of 20 MPa. For hardness, the same effects can be noticed. The most heterogeneous are the H0.27CNT and m0.4CNT coatings, due either to appearing agglomerates or variable layer thickness, or both. The 0.27CNT specimen's nanohardness and Young's modulus distribution graphs show "uplifts," which could result from the probe's contact with the surface of the native material. Nevertheless, the 0.27CNT sample appears to possess a Young's modulus very similar to that of a human bone. Cuppone et al. reported that the cortical bone has an average Young's modulus value of 18.6 ± 1.9 GPa [47]. In conclusion, the results generally show that Young's modulus increases with rising nanohardness.

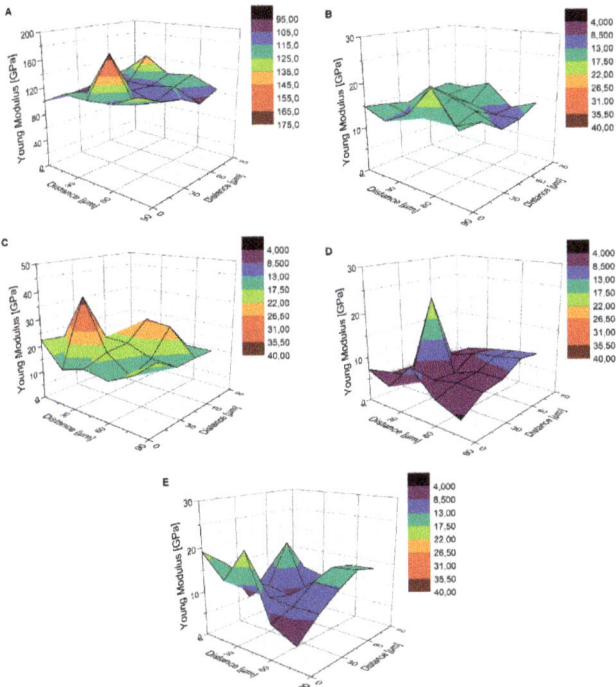

Figure 4. 3D Young's modulus distribution for: (**A**) native material after grinding (MR), (**B**) native material after etching (MRe), (**C**) CNTs coating (0.27CNT), (**D**) CNTs deposited on HAp coating (H0.27CNT), (**E**) mixed coating consisting of CNTs, nanoHAp, nanoAg and nanoCu (m0.4CNT).

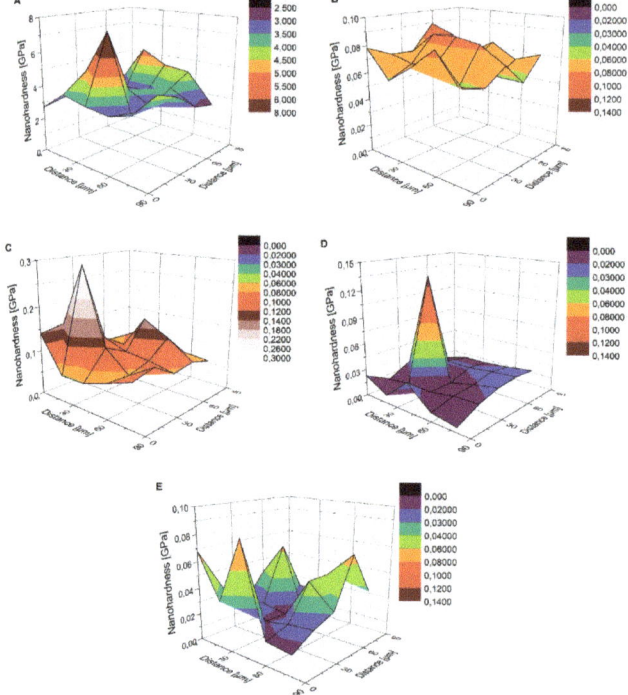

Figure 5. 3D nanohardness distribution for (**A**) native material after grinding (MR), (**B**) native material after etching (MRe), (**C**) CNTs coating (0.27CNT), (**D**) CNTs deposited on HAp coating (H0.27CNT), (**E**) mixed coating consisting of CNTs, nanHAp, nanoAg and nanoCu (m0.4CNT).

3.2.2. Nanoscratch Test

Figure 6 shows the relation of friction (F) on load (L) for each coating subjected to the nanoscratch test. The graphs describe a critical load (Lc) for every single measurement, represented by a vertical line indicating the moment of delamination for the carbon nanotube coating (0.27CNT), the carbon nanotube coating on hydroxyapatite (H0.27CNT) and the mixed coating consisting of carbon nanotubes, nanohydroxyapatite, nanosilver and nanocopper (m0.4CNT), respectively.

The values of the critical load (Lc) and critical friction (Fc), which indicate the load and friction under which the coating cracks or is delaminated, are shown in Table 5.

To conclude, the CNTs coating deposited on the Ti13Nb13Zr alloy (0.27CNT) has the best strength adhesion to the surface, while the worst adhesion is demonstrated by the composite coating (m0.4CNT). Application of HAp ceramics as an interlayer and its sintering does not improve the adhesion of the CNTs coating to the surface of titanium alloy. What is more, the addition of nanosilver and nanocopper to the composite coating further decreases adhesion, presumably due to the change in the particle size of nanometals caused by agglomeration in the bath. These results demonstrate that adhesion is best when the CNTs adhere directly to the surface, forming strong chemical bonds. Regretfully, these results cannot be compared to the adhesion strength 18–22 MPa measured by the Adhesion Test method [16] and 32 MPa measured by the F1044 shear bond strength test [35] as these methods are very different from the nanoscratch tests. On the other hand, during surgery and the period after implantation, the coatings are subject to shear stresses. Therefore, the nanoscratch method seems particularly suitable for determining real adhesion.

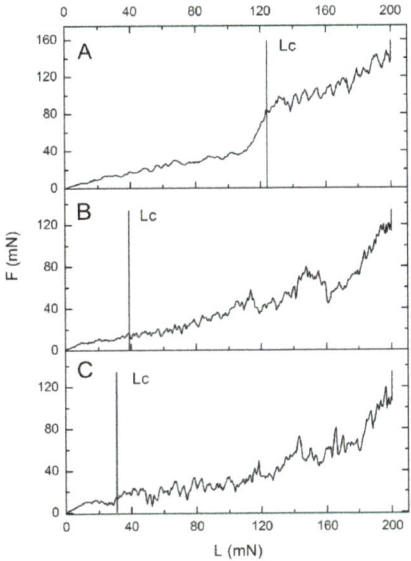

Figure 6. Friction (F) dependence on load (L) with the indicated critical load (Lc) for the single measurement of (**A**) the carbon nanotube coating on the surface of Ti13Nb13Zr (0.27CNT), (**B**) the carbon nanotube coating on the surface of the hydroxyapatite (H0.27CNT), (**C**) the mixed coating consisting of carbon nanotubes, nanohydroxyapatite, nanosilver and nanocopper (m0.4CNT).

Table 5. Parameters of coatings delamination.

Sample	Critical Friction (Fc) [mN]	Critical Load (Lc) [mN]
0.27CNT	89.42 ± 36.19	116.50 ± 32.07
H0.27CNT	39.96 ± 18.07	92.06 ± 34.3
m0.4CNT	29.56 ± 6.92	60.38 ± 10.21

3.3. Contact Angle Measurements

Figure 7 shows images of the water drops poured on the surface of all tested specimens. After etching, the surface of the Ti13Nb13Zr became hydrophobic, as the mean contact angle reached the value of 97.40°. This could be the result of the higher roughness of the etched surface (Table 3). The CNTs coating was found hydrophilic [48] with the angle value 56.50°, which means that the surface is proper for its application in implantology. Türka et al. achieved similar results for the contact angle of the CNTs after functionalization, 53.29° [49]. However, the CNTs–HAp and composite coatings are both hydrophobic, and as such they cannot be useful for implants. Prodana et al. explained the hydrophilicity of TiO_2/MWCNTs/HAp coating on titanium substrate by the appearance of C–O single bonds, and C=O and O–C=O more oxidized forms, which might be responsible for increasing surface hydrophilicity [20]. Nevertheless, during the electrophoretic deposition process, the pH rises at the cathodic substrate, resulting in deprotonation of carboxyl groups (COOH) and, thereby, functionalized CNTs become more negatively charged. Carboxyl groups (COO–) facilitate the electrostatic interaction of Ca^{2+} ions from HAp with CNTs [46]. Based on reports to date, and our research, the hydrophobic appearance of an arrangement of CNTs and porous HAp may be attributed to an occurrence of a specific architecture with a lesser ability to form van der Waals bonds between water and specimen surface. A similar explanation may be given regarding the composite materials, indicating the increased effects of nanohydroxyapatite, which may have a different microstructure to CNTs. Further attempts made with various forms of both components, oriented particularly towards better hydrophilicity, should give a more plausible solution.

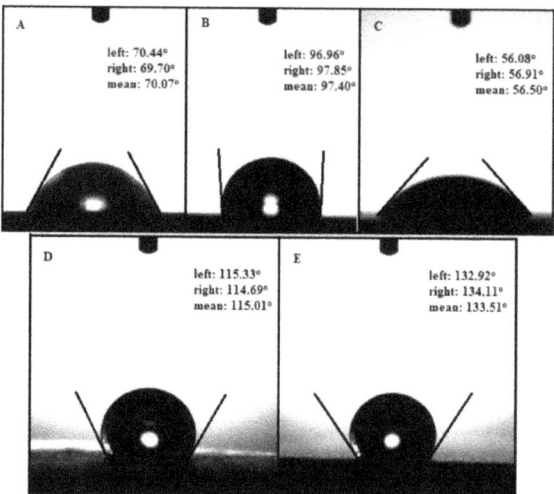

Figure 7. Contact angle (CA) for: (**A**) native material after grinding (MR), (**B**) native material after etching (MRe), (**C**) carbon nanotube coating on the surface of Ti13Nb13Zr (0.27CNT), (**D**) carbon nanotube coating on the surface of the hydroxyapatite (H0.27CNT), (**E**) the mixed coating consisting of carbon nanotubes, nanohydroxyapatite, nanosilver and nanocopper (m0.4CNT).

4. Conclusions

The multi-wall carbon nanotubes deposited on the surface of the Ti13Nb13Zr alloy by the electrophoretic method demonstrate a relatively high hardness and wear resistance, firm adhesion, and proper wettability, so they are suitable for surface treatment of the titanium implants made of the investigated alloy.

The hybrid coatings obtained by the formation of hydroxyapatite deposit, followed by a carbon nanotubes' layer, have lower adhesion strength, hardness, and less proper wettability, presumably due to the weak interface bonding between hydroxyapatite and carbon nanotubes, which is related to the properties of the ceramic material.

The composite layer has the lowest adhesion strength, hardness, and equally improper wettability, which can be attributed to weak bonding of components in the coating.

The obtained results provide evidence that the mutual bonding strength of all components and critical strength at the coating–substrate interface, are more significant determinants of mechanical properties than the components and the deposition process parameters.

Author Contributions: Conceptualization, B.M.-M.; Formal Analysis, M.B. and B.B.; Investigation, B.M.-M., D.R.-W. and M.B.; Methodology, B.M.-M., D.R.-W., M.B. and B.B.; Project administration, B.M.-M.; Resources, B.B.; Supervision, A.Z.; Visualization, D.R.-W.; Writing–Original Draft, B.M.-M. and D.R.-W.; Writing–Review and Editing, A.Z.

Funding: This research received no external funding.

Acknowledgments: We are grateful to Julia Mazurowska for her help in carrying out experiments and Bartłomiej Jankiewicz's team from the Military University of Technology, Institute of Optoelectronics for sharing laboratory equipment and support in preparing materials to tests.

Conflicts of Interest: The authors declare no conflict of interest.

References

1. Ku, S.H.; Lee, M.; Park, C.B. Carbon-Based Nanomaterials for Tissue Engineering. *Adv. Healthc. Mater.* **2013**, *2*, 244–260. [CrossRef]
2. Li, X.; Liu, X.; Huang, J.; Fan, Y.; Cui, F. Biomedical investigation of CNT based coatings. *Surf. Coat. Technol.* **2011**, *206*, 759–766.
3. Dlugon, E.; Simka, W.; Fraczek-Szczypta, A.; Niemiec, W.; Markowski, J.; Szymanska, M.; Blazewicz, M. Carbon nanotube-based coatings on titanium. *Bull. Mater. Sci.* **2015**, *38*, 1339–1344. [CrossRef]
4. Tanaka, M.; Sato, Y.; Zhang, M.; Haniu, H.; Okamoto, M.; Aoki, K.; Takizawa, T.; Yoshida, K.; Sobajima, A.; Kamanaka, T.; et al. In Vitro and In Vivo Evaluation of a Three-Dimensional Porous Multi-Walled Carbon Nanotube Scaffold for Bone Regeneration. *Nanomaterials* **2017**, *7*, 46. [CrossRef] [PubMed]
5. Lahiri, D.; Ghosh, S.; Agarwal, A. Carbon nanotube reinforced hydroxyapatite composite for orthopedic application: A review. *Mater. Sci. Eng. C* **2012**, *32*, 1727–1758. [CrossRef]
6. Usui, Y.; Aoki, K.; Narita, N.; Murakami, N.; Nakamura, I.; Nakamura, K.; Ishigaki, N.; Yamazaki, H.; Horiuchi, H.; Kato, H.; et al. Carbon Nanotubes with High Bone-Tissue Compatibility and Bone-Formation Acceleration Effects. *Small* **2008**, *8621*, 240–246. [CrossRef] [PubMed]
7. Kalbacova, M.; Kalbac, M. Influence of single-walled carbon nanotube films on metabolic activity and adherence of human osteoblasts. *Carbon* **2007**, *45*, 2266–2272. [CrossRef]
8. Lahiri, D.; Benaduce, A.P.; Rouzaud, F.; Solomon, J.; Keshri, A.K.; Kos, L.; Agarwal, A. Wear behavior and in vitro cytotoxicity of wear debris generated from hydroxyapatite–carbon nanotube composite coating. *J. Biomed. Mater. Res. Part A* **2010**, *0494*, 1–12. [CrossRef]
9. Matsuoka, M.; Akasaka, T.; Totsuka, Y.; Watari, F. Strong adhesion of Saos-2 cells to multi-walled carbon nanotubes. *Mater. Sci. Eng. B* **2010**, *173*, 182–186. [CrossRef]
10. Akasaka, T.; Yokoyama, A.; Matsuoka, M.; Hashimoto, T.; Watari, F. Thin films of single-walled carbon nanotubes promote human osteoblastic cells (Saos-2) proliferation in low serum concentrations. *Mater. Sci. Eng. C* **2010**, *30*, 391–399. [CrossRef]
11. Hirschfeld, J.; Akinoglu, E.M.; Wirtz, D.C.; Hoerauf, A.; Bekeredjian-Ding, I.; Jepsen, S.; Haddouti, E.M.; Limmer, A.; Giersig, M. Long-term release of antibiotics by carbon nanotube-coated titanium alloy surfaces diminish biofilm formation by Staphylococcus epidermidis. *Nanomed. Nanotechnol. Biol. Med.* **2017**, *13*, 1587–1593. [CrossRef] [PubMed]
12. Bai, Y.; Prasad, M.; Song, I.; Ho, M.; Sung, T.; Watari, F.; Uo, M. Electrophoretic deposition of carbon nanotubes – hydroxyapatite nanocomposites on titanium substrate. *Mater. Sci. Eng. C* **2010**, *30*, 1043–1049. [CrossRef]
13. Długoń, E.; Niemiec, W.; Fraczek-Szczypta, A.; Jeleń, P.; Sitarz, M.; Błazewicz, M. Spectroscopic studies of electrophoretically deposited hybrid HAp/CNT coatings on titanium. *Spectrochim. Acta Part A Mol. Biomol. Spectrosc.* **2014**, *133*, 872–875. [CrossRef]
14. Abrishamchian, A.; Hooshmand, T.; Mohammadi, M.; Najafi, F. Preparation and characterization of multi-walled carbon nanotube/hydroxyapatite nanocomposite film dip coated on Ti-6Al-4V by sol-gel method for biomedical applications: An in vitro study. *Mater. Sci. Eng. C* **2013**, *33*, 2002–2010. [CrossRef]
15. Gopi, D.; Shinyjoy, E.; Kavitha, L. Influence of ionic substitution in improving the biological property of carbon nanotubes reinforced hydroxyapatite composite coating on titanium for orthopedic applications. *Ceram. Int.* **2015**, *41*, 5454–5463. [CrossRef]
16. Gopi, D.; Shinyjoy, E.; Sekar, M.; Surendiran, M.; Kavitha, L.; Sampath Kumar, T.S. Development of carbon nanotubes reinforced hydroxyapatite composite coatings on titanium by electrodeposition method. *Corros. Sci.* **2013**, *73*, 321–330. [CrossRef]
17. Pei, X.; Zeng, Y.; He, R.; Li, Z.; Tian, L.; Wang, J.; Wan, Q.; Li, X.; Bao, H. Single-walled carbon nanotubes/hydroxyapatite coatings on titanium obtained by electrochemical deposition. *Appl. Surf. Sci.* **2014**, *295*, 71–80. [CrossRef]
18. Zhong, Z.; Qin, J.; Ma, J. Electrophoretic deposition of biomimetic zinc substituted hydroxyapatite coatings with chitosan and carbon nanotubes on titanium. *Ceram. Int.* **2015**, *41*, 8878–8884. [CrossRef]
19. Zanello, L.P.; Zhao, B.; Hu, H.; Haddon, R.C. Bone cell proliferation on carbon nanotubes. *Nano Lett.* **2006**, *6*, 562–567. [CrossRef]

20. Prodana, M.; Duta, M.; Ionita, D.; Bojin, D.; Stan, M.S.; Dinischiotu, A.; Demetrescu, I. A new complex ceramic coating with carbon nanotubes, hydroxyapatite and TiO 2 nanotubes on Ti surface for biomedical applications. *Ceram. Int.* **2015**, *41*, 6318–6325. [CrossRef]
21. Chouirfa, H.; Bouloussa, H.; Migonney, V.; Falentin-Daudré, C. Review of titanium surface modification techniques and coatings for antibacterial applications. *Acta Biomater.* **2019**, *83*, 37–54. [CrossRef] [PubMed]
22. Sivaraj, D.; Vijayalakshmi, K. Substantial effect of magnesium incorporation on hydroxyapatite/carbon nanotubes coatings on metallic implant surfaces for better anticorrosive protection and antibacterial ability. *J. Anal. Appl. Pyrolysis* **2018**, *135*, 15–21. [CrossRef]
23. Mukherjee, S.; Nandi, S.K.; Kundu, B.; Chanda, A.; Sen, S.; Das, P.K. Enhanced bone regeneration with carbon nanotube reinforced hydroxyapatite in animal model. *J. Mech. Behav. Biomed. Mater.* **2016**, *60*, 243–255. [CrossRef] [PubMed]
24. Malik, M.A.; Wani, M.Y.; Hashim, M.A.; Nabi, F. Nanotoxicity: Dimensional and morphological concerns. *Adv. Phys. Chem.* **2011**, *2011*.
25. Teleanu, D.; Chircov, C.; Grumezescu, A.; Teleanu, R. Neurotoxicity of Nanomaterials: An Up-to-Date Overview. *Nanomaterials* **2019**, *9*, 96. [CrossRef]
26. Mohanta, D.; Patnaik, S.; Sood, S.; Das, N. Carbon nanotubes: Evaluation of toxicity at biointerfaces. *J. Pharm. Anal.* **2019**. [CrossRef]
27. Nawrotek, K.; Tylman, M.; Rudnicka, K.; Gatkowska, J.; Balcerzak, J. Tubular electrodeposition of chitosan-carbon nanotube implants enriched with calcium ions. *J. Mech. Behav. Biomed. Mater.* **2016**, *60*, 256–266. [CrossRef]
28. Guan, K.; Zhang, L.; Zhu, F.; Sheng, H.; Li, H. Surface modification for carbon/carbon composites with Mg-CaP coating reinforced by SiC nanowire-carbon nanotube hybrid for biological application. *Appl. Surf. Sci.* **2019**, *489*, 856–866. [CrossRef]
29. Khazeni, D.; Saremi, M.; Soltani, R. Development of HA-CNTs composite coating on AZ31 Magnesium alloy by cathodic electrodeposition. Part 2: Electrochemical and in-vitro behavior. *Ceram. Int.* **2019**, *45*, 11186–11194. [CrossRef]
30. Alsagri, A.S.; Nasir, S.; Gul, T.; Islam, S.; Nisar, K.S.; Shah, Z.; Khan, I. MHD thin film flow and thermal analysis of blood with CNTs nanofluid. *Coatings* **2019**, *9*, 175. [CrossRef]
31. Przekora, A.; Benko, A.; Nocun, M.; Wyrwa, J.; Blazewicz, M.; Ginalska, G. Titanium coated with functionalized carbon nanotubes—A promising novel material for biomedical application as an implantable orthopaedic electronic device. *Mater. Sci. Eng. C* **2014**, *45*, 287–296. [CrossRef] [PubMed]
32. Jacobs, C.B.; Peairs, M.J.; Venton, B.J. Carbon nanotube based electrochemical sensors for biomolecules. *Anal. Chim. Acta* **2010**, *662*, 105–127. [CrossRef] [PubMed]
33. Benko, A.; Nocuń, M.; Berent, K.; Gajewska, M.; Klita, Ł.; Wyrwa, J.; Błażewicz, M. Diluent changes the physicochemical and electrochemical properties of the electrophoretically-deposited layers of carbon nanotubes. *Appl. Surf. Sci.* **2017**, *403*, 206–217. [CrossRef]
34. Sivaraj, D.; Vijayalakshmi, K. Novel synthesis of bioactive hydroxyapatite/f-multiwalled carbon nanotube composite coating on 316L SS implant for substantial corrosion resistance and antibacterial activity. *J. Alloys Compd.* **2019**, 1340–1346. [CrossRef]
35. Maleki-Ghaleh, H.; Khalil-Allafi, J. Characterization, mechanical and in vitro biological behavior of hydroxyapatite-titanium-carbon nanotube composite coatings deposited on NiTi alloy by electrophoretic deposition. *Surf. Coatings Technol.* **2019**, *363*, 179–190. [CrossRef]
36. Mohajernia, S.; Pour-Ali, S.; Hejazi, S.; Saremi, M.; Kiani-Rashid, A.R. Hydroxyapatite coating containing multi-walled carbon nanotubes on AZ31 magnesium: Mechanical-electrochemical degradation in a physiological environment. *Ceram. Int.* **2018**, *44*, 8297–8305. [CrossRef]
37. Park, J.E.; Jang, Y.S.; Bae, T.S.; Lee, M.H. Multi-walled carbon nanotube coating on alkali treated TiO2 nanotubes surface for improvement of biocompatibility. *Coatings* **2018**, *8*, 159. [CrossRef]
38. Fraczek-Szczypta, A.; Jantas, D.; Ciepiela, F.; Grzonka, J.; Bernasik, A.; Marzec, M. Carbon nanomaterials coatings—Properties and influence on nerve cells response. *Diam. Relat. Mater.* **2018**, *84*, 127–140.
39. Farrokhi-Rad, M.; Menon, M. Effect of Dispersants on the Electrophoretic Deposition of Hydroxyapatite-Carbon Nanotubes Nanocomposite Coatings. *J. Am. Ceram. Soc.* **2016**, *99*, 2947–2955. [CrossRef]

40. Liu, S.; Li, H.; Su, Y.; Guo, Q.; Zhang, L. Preparation and properties of in-situ growth of carbon nanotubes reinforced hydroxyapatite coating for carbon/carbon composites. *Mater. Sci. Eng. C* **2017**, *70*, 805–811. [CrossRef]
41. Singh, I.; Kaya, C.; Shaffer, M.S.; Thomas, B.C.; Boccaccini, A.R. Bioactive ceramic coatings containing carbon nanotubes on metallic substrates by electrophoretic deposition. *J. Mater. Sci.* **2006**, *41*, 8144–8151. [CrossRef]
42. Constanda, S.; Stan, M.S.; Ciobanu, C.S.; Motelica-Heino, M.; Guégan, R.; Lafdi, K.; Dinischiotu, A.; Predoi, D. Carbon Nanotubes-Hydroxyapatite Nanocomposites for an Improved Osteoblast Cell Response. *J. Nanomater.* **2016**, *2016*. [CrossRef]
43. Yi, C.; Bagchi, S.; Dmuchowski, C.M.; Gou, F.; Chen, X.; Park, C.; Chew, H.B.; Ke, C. Direct nanomechanical characterization of carbon nanotube - titanium interfaces. *Carbon N. Y.* **2018**, *132*, 548–555. [CrossRef]
44. Sasani, N.; Vahdati Khaki, J.; Mojtaba Zebarjad, S. Characterization and nanomechanical properties of novel dental implant coatings containing copper decorated-carbon nanotubes. *J. Mech. Behav. Biomed. Mater.* **2014**, *37*, 125–132. [CrossRef]
45. Zhao, X.; Chen, X.; Zhang, L.; Liu, Q.; Wang, Y.; Zhang, W.; Zheng, J. Preparation of Nano-Hydroxyapatite Coated Carbon Nanotube Reinforced Hydroxyapatite Composites. *Coatings* **2018**, *8*, 357. [CrossRef]
46. Fathyunes, L.; Khalil-Allafi, J.; Moosavifar, M. Development of graphene oxide/calcium phosphate coating by pulse electrodeposition on anodized titanium: Biocorrosion and mechanical behavior. *J. Mech. Behav. Biomed. Mater.* **2019**, *90*, 575–586. [CrossRef]
47. Cuppone, M.; Seedhom, B.B.; Berry, E.; Ostell, A.E. The Longitudinal Young's Modulus of Cortical Bone in the Midshaft of Human Femur and its Correlation with CT Scanning Data. *Calcif. Tissue Int.* **2004**, *74*, 302–309.
48. Sansotera, M.; Talaeemashhadi, S.; Gambarotti, C.; Pirola, C.; Longhi, M.; Ortenzi, M.A.; Navarrini, W.; Bianchi, C.L. Comparison of branched and linear perfluoropolyether chains functionalization on hydrophobic, morphological and conductive properties of multi-walled carbon nanotubes. *Nanomaterials* **2018**, *8*, 176. [CrossRef]
49. Türk, S.; Altınsoy, I.; Çelebi Efe, G.; Ipek, M.; Özacar, M.; Bindal, C. 3D porous collagen/functionalized multiwalled carbon nanotube/chitosan/hydroxyapatite composite scaffolds for bone tissue engineering. *Mater. Sci. Eng. C* **2018**, *92*, 757–768. [CrossRef]

© 2019 by the authors. Licensee MDPI, Basel, Switzerland. This article is an open access article distributed under the terms and conditions of the Creative Commons Attribution (CC BY) license (http://creativecommons.org/licenses/by/4.0/).

Article

Electrophoretic Deposition and Characterization of Chitosan/Eudragit E 100 Coatings on Titanium Substrate

Łukasz Pawłowski [1,*], Michał Bartmański [1], Gabriel Strugała [1], Aleksandra Mielewczyk-Gryń [2], Magdalena Jażdżewska [1] and Andrzej Zieliński [1]

[1] Faculty of Mechanical Engineering, Gdańsk University of Technology, Narutowicza 11/12, 80-233 Gdańsk, Poland; michal.bartmanski@pg.edu.pl (M.B.); gabriel.strugala@pg.edu.pl (G.S.); magdalena.jazdzewska@pg.edu.pl (M.J.); andrzej.zielinski@pg.edu.pl (A.Z.)

[2] Faculty of Applied Physics and Mathematics, Gdańsk University of Technology, Narutowicza 11/12, 80-233 Gdańsk, Poland; alegryn@pg.edu.pl

* Correspondence: lukasz.pawlowski@pg.edu.pl; Tel.: +48-883-797-081

Received: 29 May 2020; Accepted: 26 June 2020; Published: 28 June 2020

Abstract: Currently, a significant problem is the production of coatings for titanium implants, which will be characterized by mechanical properties comparable to those of a human bone, high corrosion resistance, and low degradation rate in the body fluids. This paper aims to describe the properties of novel chitosan/Eudragit E 100 (chit/EE100) coatings deposited on titanium grade 2 substrate by the electrophoretic technique (EPD). The deposition was carried out for different parameters like the content of EE100, time of deposition, and applied voltage. The microstructure, surface roughness, chemical and phase composition, wettability, mechanical and electrochemical properties, and degradation rate at different pH were examined in comparison to chitosan coating without the addition of Eudragit E 100. The applied deposition parameters significantly influenced the morphology of the coatings. The chit/EE100 coating with the highest homogeneity was obtained for Eudragit content of 0.25 g, at 10 V, and for 1 min. Young's modulus of this sample (24.77 ± 5.50 GPa) was most comparable to that of human cortical bone. The introduction of Eudragit E 100 into chitosan coatings significantly reduced their degradation rate in artificial saliva at neutral pH while maintaining high sensitivity to pH changes. The chit/EE100 coatings showed a slightly lower corrosion resistance compared to the chitosan coating, however, significantly exceeding the substrate corrosion resistance. All prepared coatings were characterized by hydrophilicity.

Keywords: titanium; chitosan; Eudragit; electrophoretic deposition; nanoindentation; pH-sensitive coatings; wettability

1. Introduction

Titanium and titanium alloys are materials often used in biomedical applications due to their high biocompatibility, high corrosion resistance, and low Young's modulus comparing to other metallic biomaterials. These properties promote their use as orthopedic and dental implants, orthodontic wires and brackets, and other biomedical devices [1–3]. Titanium and its alloys are often subjected to improving their osseointegration properties, resistance to corrosion, and protection against the development of bacterial infections by modification of the surface topography and the deposition of bioactive materials, e.g., calcium phosphates and bioglasses [4–7].

Currently, so-called smart polymers that respond to the external environment are gathering considerable interest. These materials change their properties under the influence of temperature, pH, UV–Vis radiation, electric, and magnetic field effects [8–12]. Among the most popular polymers are chitosan [13] and Eudragit E 100 (EE100)–methacrylic acid copolymer [14].

Chitosan, due to its biodegradability, biocompatibility, nontoxicity, and antibacterial activity, is often used in controlled drug delivery systems, wound healing, and tissue regeneration [15,16]. It is commonly applied in different forms, such as membranes, nanogels, micro/nanoparticles, films, and hydrogels [17–19]. Chitosan coatings are gathering more interest in implantology, however, they show low mechanical properties and low stability at neutral pH [20,21]. Chitosan rapidly absorbs water and is characterized by a high swelling degree in aqueous environments, leading to fast drug release [22]. Hence, co-deposition of chitosan with other biopolymers (e.g., gelatin) or nanomaterials (e.g., silver nanoparticles, gold nanoparticles, carbon nanotubes) is performed to overcome these problems [23–25]. One of the modifiers of chitosan coatings may be EE100.

Eudragits are a group of biopolymer materials that have been used in controlled drug delivery systems for several years. Depending on their functional groups, they are usually divided into polycations and polyanions [26]. Polycations include Eudragits E with dimethylamino groups, and RL, ES, NE with quaternary amino groups, while polyanions include Eudragits L and S with carboxyl groups [27]. Eudragit E 100 copolymer is based on dimethylaminoethyl methacrylate, butyl methacrylate, and methyl methacrylate with a ratio of 2:1:1 and belongs to the group of cationic polymers. It is sensitive to pH changes, dissolves in acidic environments due to amino groups, but is insoluble at neutral pH [28]. In an alkaline environment, the polymer swells [29]. It is most often used for coatings on pills to transport the drug substance to the appropriate part of the digestive tract, mainly the stomach, because of its good adhesion, low viscosity, and good ability to mask odor and unpleasant taste [30]. The sensitivity of Eudragit E 100 to change in pH is utilized in drug delivery systems because inflamed and cancerous tissues are characterized by a lower pH value [31,32]. This copolymer is mainly used as a coating material, nanocapsules, or nanoparticles. EE100 is also widely used to improve the solubility of drugs that are poorly soluble in water. The drug substance is then either dispersed in the biopolymer matrix or trapped in the nanocapsules [28,33–41].

Eudragit E 100 is often used as a blend with other biopolymers, which can result in the development of a new biopolymer with desired properties, such as a drug release profile. Farooq et al. produced Eudragit E 100/polycaprolactone microspheres in oil by the water solvent evaporation method [30]. Blended polylactic glycolic acid and Eudragit E 100 were proposed for the prevention of autoimmune diabetes [42]. There are many reports of the use of Eudragit E 100 in medicine, but few of them relate to bone implants.

There are several reports on the use of chitosan in combination with Eudragits. Vibhooti et al. developed Eudragit S 100-coated chitosan beads with pH-sensitivity for colon-targeted delivery [43]. Eudragit L 100 and S 100 were used for coating crosslinked chitosan microspheres with metronidazole by the emulsion solvent evaporation technique [44]. Interpolyelectrolyte complexes of chitosan and Eudragit L 100 were applied in oral controlled drug delivery systems [45]. Xu et al. prepared Eudragit L 100-coated mannosylated chitosan nanoparticles for oral bovine serum albumin delivery [46]. The addition of Eudragit RS to the pectin/chitosan films prepared by the casting/solvent evaporation method significantly decreased the swelling ratio of this polyelectrolyte complex in phosphate-buffered saline (PBS). Furthermore, the introduction of the Eudragit RS to the coating ensured a controllable slow release followed by a burst release of theophylline immediately after the change in pH [47]. Kouchak et al. revealed that increasing Eudragit's RL content in chitosan films could improve their mechanical properties without undesirable effects on their water uptake and oxygen penetration. The properties of these biopolymer coatings can be modified by changing the chitosan/Eudragit ratio [48].

The aim of this research is the electrophoretic deposition and characterization of the chit/EE100 coatings. According to the previous studies [47,48], the addition of Eudragit E 100 improves the mechanical properties of chitosan coatings and limits the dissolution rate of the chitosan coating at neutral pH. This type of coating may be a matrix for the controlled release of the drug used in the case of load-bearing implants. So far, there have been no reports in the literature regarding the production of this type of biopolymer coatings using the electrophoretic deposition method.

2. Materials and Methods

2.1. Materials

The Ti grade 2 (EkspresStal, Luboń, Poland) was used as a substrate. Table 1 shows its chemical composition given by the manufacturer. Commercial high molecular weight chitosan (high purity > 99%, MW ~ 310–375 kDa) coarse ground flakes and powder with a degree of deacetylation > 75% were purchased from Sigma-Aldrich (St. Louis, MO, USA). Eudragit E 100 granules (purity 99.9%, MW ~ 47 kDa) were provided by the Evonik Industries (Darmstadt, Germany). Acetic acid (99.9%) was obtained from Stanlab (Gliwice, Poland), while isopropanol (99.8%) and hydrochloric acid (30%) from POCH (Gliwice, Poland).

Table 1. The chemical composition of the Ti grade 2 substrate, wt.%.

Element	N	C	H	Fe	O	Ti
wt.%	0.009	0.013	0.001	0.168–0.179	0.190–0.170	remainder

2.2. Substrate Preparation

As a substrate, the Ti grade 2 round samples with a diameter of 12 mm and a height of 4 mm (cut from a rod) were used. All samples were wet ground using SiC abrasive papers up to grit #800. Prior to coating deposition, the Ti substrate was rinsed with isopropanol and distilled water.

2.3. Electrophoretic Deposition of Chitosan/Eudragit E 100 Coatings

Two different suspensions containing 0.25 g (suspension A) and 0.5 g (suspension B) of Eudragit E 100 were prepared for electrophoretic deposition. The appropriate amount of biopolymer was dissolved in 100 mL of 1% (v/v) aqueous acetic acid solution with 0.1 g of chitosan, according to the previous work [49]. This suspension was magnetically stirred (Dragon Lab MS-H-Pro+, Schiltigheim, France) for 24 h at room temperature.

Different time and deposition voltage values were used. The designation of samples with applied parameters is shown in Table 2. Ti substrate was used as a cathode, and the counter electrode was platinum mesh. The distance between electrodes connected to the DC power source (MCP/SPN110-01C, Shanghai MCP Corp., Shanghai, China) was about 10 mm. The deposition was carried out at room temperature. After deposition, the samples were rinsed with distilled water, dried at room temperature for 48 h, and stored in a desiccator for further characterization. After the deposition, the parameters ensuring the best quality of chit/EE100 coating were selected, and, for comparison, a chitosan coating without the addition of EE100 was prepared using these deposition parameters.

Table 2. Designations of experiment samples with the applied process parameters.

Suspension		Sample	Voltage (V)	Time (min)
A (0.25 g EE100)	100 mL of 1% (v/v) acetic acid with 0.1 g of chitosan	A1 A3	10	1 3
		A1' A3'	30	1 3
B (0.5 g EE100)		B1 B3	10	1 3
		B1' B3'	30	1 3

2.4. Structure and Morphology of Chitosan/Eudragit E 100 Coatings

The surfaces of the composite coating were examined using a high resolution scanning electron microscope (SEM JEOL JSM-7800 F, JEOL Ltd., Tokyo, Japan) with an LED detector at 5 kV acceleration voltage. Before testing, samples were sputtered with a 10 nm thick layer of gold using a table-top DC magnetron sputtering coater (EM SCD 500, Leica, Vienna, Austria) in a pure Ar plasma condition (Argon, Air Products 99.999%). The surface roughness of all prepared samples was determined by using a contact profilometer with EVOVIS software (1.38.0.2) (Hommel Etamic Waveline, Jenoptik, Jena, Germany). The test was conducted according to the ISO 4287-1997 standard [50]. Three measurements were carried out for each sample, measurement distance was 8.8 mm with a scanning speed of 0.5 mm/s. Based on the tests, the average values of roughness (Ra), the peak-to-valley roughness (Rz), and maximum peak-to-mean height (Rp) were obtained. The qualitative elemental analysis of the obtained coatings was determined by the X-ray energy-dispersive spectrometer (EDS) (Edax Inc., Mahwah, NJ, USA). The X-ray diffraction spectroscopy (Phillips X'Pert Pro, Almelo, the Netherlands) was conducted (Cu Kα, λ = 0.1554 nm) in the 2θ range of 10°–90° at a 0.02 step and 2 s/point at ambient temperature and under atmospheric pressure. Fourier-transform infrared spectroscopy (FTIR, Perkin Elmer Frontier, Waltham, MA, USA) at a resolution of 2 cm^{-1} (scans number 32) in the range of 400–4000 cm^{-1} was utilized.

2.5. Mechanical Studies

Nanoindentation tests were performed using the NanoTest™ Vantage device (Micro Materials, Wrexham, Great Britain) with a Berkovich three-sided pyramidal diamond indenter. Ten independent measurements were performed for the Ti reference sample and the biopolymer coatings prepared at different deposition parameters. The distance between individual indents was 20 µm. The value of maximum force was 50 mN, the loading and unloading rate were set up at 20 s and the dwell period at maximum load was 10 s. The load–displacement curve was obtained for each measurement by the Oliver and Pharr method. Based on these curves, surface hardness (H) and Young's modulus (E) were determined. For the calculations, the values of Poisson's ratio 0.3 and 0.4 were used for the reference Ti sample and the samples with biopolymer coatings, respectively.

The scratch tests were carried out over a distance of 500 µm, the load increasing from 0 to 200 mN at a loading rate of 1.3 mN/s. The force that caused complete delamination of the coating from the substrate was determined based on an abrupt change in frictional force at the plot of the normal force versus friction force for each measurement. Besides, for its exact determination, all scratches were examined using an optical microscope (BX51, OLYMPUS, Tokyo, Japan).

2.6. Degradation Analysis

Dried and pre-weighed (Pioneer PA114CM/1, OHAUS, Greifensee, Switzerland) samples with chitosan and chit/EE100 coatings were immersed in artificial saliva solution (ASS, prepared according to reference [51]) at 37 °C temperature at different pH (3, 5, and 7) value for 1, 3 and 7 days. HCl was used to adjust the solution pH. According to reference [52], weight loss (WL) of the investigated coating was calculated as:

$$WL = \frac{W_1 - W_2}{W_1} \times 100\% \tag{1}$$

where W_1 is the weight of the dry sample with coating before swelling and W_2 is the weight of the dry sample after swelling. The measurement results were collected at an accuracy of 0.0001 g.

2.7. Corrosion Studies

The electrochemical measurements were made in a potentiodynamic mode in artificial saliva solution at 37 °C using a potentiostat/galvanostat (Atlas 0531, Atlas Sollich, Gdansk, Poland). A three-electrode cell setup was utilized, with platinum electrode as a counter electrode, and Ag/AgCl (saturated with potassium chloride) as a reference electrode. Before the experiment, the samples were stabilized at their open circuit potential (OCP) for 10 min. A potentiodynamic polarization test was conducted within a scan range −800/1000 mV at a potential change rate of 1 mV/s. Using the Tafel extrapolation method, the corrosion potential (E_{corr}) and corrosion current density (i_{corr}) values were determined.

2.8. Contact Angle Studies

The measurement of the water contact angle was carried out by falling drop method (Contact Angle Goniometer, Zeiss, Oberkochen, Germany) at room temperature and 10 s after drop out. The water drop volume was about 2 µL, and three measurements were performed for each sample.

3. Results and Discussion

3.1. Structure and Morphology of Chitosan/Eudragit E 100 Coatings

Figure 1 depicts the microstructure of the Ti grade 2 substrate, the chitosan coating, and the chit/EE100 coatings obtained by electrophoretic deposition. The Ti grade 2 substrate after wet grinding was characterized by a typical structure resulting from the grinding process [53]. The effects of EE100 content in the suspension, deposition time, and applied voltage on the quality of prepared coatings are visible. The increasing deposition time and the applied voltage resulted in more uneven coatings morphology. This effect has also been observed in other studies [54]. The increase in these parameters caused more rapid kinetics of coating deposition and bubble formation of hydrogen gas on the surface of the titanium sample caused by water electrolysis, which resulted in the deposition of a more heterogeneous coating [55,56]. The presence of hydrogen bubbles blocks the flow of biopolymer particles to the surface, which strongly affects the structure of coatings [57]. The prints of formed hydrogen bubbles are visible in the SEM images (Figure 1). In some areas of the coatings, it caused total exposure of the titanium substrate. The reduction in bubble formation can be achieved by reducing water content in the suspension by replacing it with, e.g., ethanol [58].

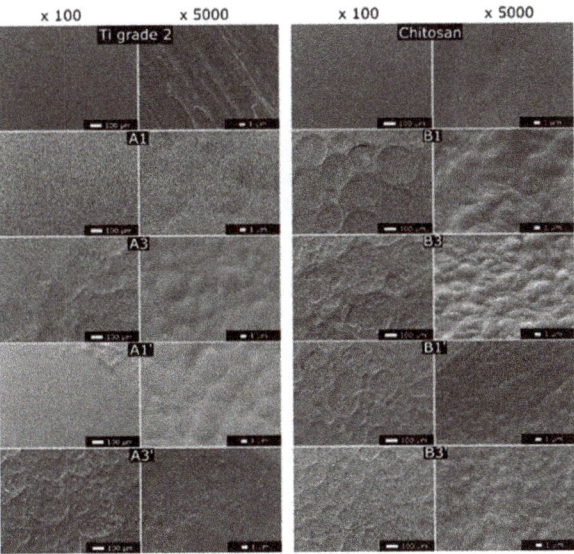

Figure 1. SEM images of the surface topography of the Ti grade 2 substrate, the chitosan coating, and the chit/EE100 coatings obtained at different deposition parameters; the images obtained at different magnifications, ×100 (on the left) and ×5000 (on the right).

An increase in the content of EE100 in the suspension also contributed to the increase in heterogeneity of the obtained coatings. Similar to other biopolymer coatings, an increase in Eudragit content in the suspension caused a disturbance in particle flow, resulting in a more porous coating [59]. For all chit/EE100 samples, the images obtained at higher magnifications showed a microporous structure of the coatings. However, it has been reported that the porosity of implant coatings promotes in vivo cell growth [23]. The A1 sample, prepared at the lowest deposition parameters, showed the highest homogeneity. In this case, the coating completely covered the titanium substrate surface, and the images obtained at higher magnifications revealed slight unevenness. Therefore, the A1 sample was selected for the next examinations. For comparison purposes, using the A1 sample deposition parameters, a chitosan coating without the addition of EE100 was prepared for the remaining tests. Continuous coating was observed, and it was characterized by uniformly distributed unevenness visible at higher magnifications. Compared to the A1 sample, a similar homogeneity was observed. The mechanism of creating chitosan coatings most likely involves loss of charge in the high pH value, an alkaline region on the cathode surface by chitosan protonated amino groups, and the formation of insoluble precipitates [60].

Table 3 summarizes the mean values of surface roughness parameters: the average roughness (Ra), the peak-to-valley roughness (Rz), and maximum peak-to-mean height (Rp) obtained for the Ti grade 2 substrate, the chitosan coating, and the chit/EE100 coatings.

Table 3. Surface roughness parameters of the Ti grade 2 substrate, the chitosan coating, and the chit/EE100 coatings (mean ± SD; n = 3).

Sample	Surface Roughness Parameters (µm)		
	Ra	Rz	Rp
Ti grade 2	0.12 ± 0.01	0.77 ± 0.13	0.44 ± 0.11
Chitosan	0.15 ± 0.05	1.34 ± 0.58	0.88 ± 0.46
A1	1.57 ± 0.05	7.01 ± 0.44	4.29 ± 0.29
A3	2.84 ± 0.11	11.39 ± 0.05	6.31 ± 0.11
A1'	4.63 ± 0.68	20.27 ± 2.05	10.24 ± 1.00
A3'	2.98 ± 0.24	14.48 ± 0.56	8.01 ± 0.37
B1	2.53 ± 0.47	12.16 ± 1.97	7.64 ± 1.65
B3	2.66 ± 0.31	12.47 ± 1.29	7.70 ± 1.24
B1'	2.39 ± 0.11	11.76 ± 0.64	7.28 ± 0.71
B3'	2.93 ± 0.38	12.92 ± 1.28	7.19 ± 0.88

The titanium substrate showed the lowest roughness. The deposition of biopolymer coatings by the electrophoretic method resulted in increased surface roughness compared to a bare substrate. A similar relationship was observed in previous studies [49]. The chitosan coating showed roughness similar to the substrate after grinding; in the case of chit/EE100 coatings, a significant increase in mean values of parameters Ra, Rz, and Rp was observed. The reason for this lies in the more rapid EPD process for chit/EE100 deposition and the formation of hydrogen bubbles on the cathode [55]. The results obtained are consistent with the SEM images shown in Figure 1. Increased surface roughness allows for better tissue adhesion and stabilization of the implant in the initial phase [61].

Figure 2 presents the results of the EDS measurements for the Ti grade 2 substrate, the sample with chitosan coating, and the sample with chit/EE100 coating (sample A1). This analysis was only qualitative. The samples previously subjected to SEM examinations were used; hence, peaks referring to Au were visible in all spectra. EDS spectrum of the substrate confirmed the presence of Ti. For the other two samples, peaks related to Ti were less intense. Moreover, constituents of the coatings (O, C) and Ti element from substrate were noted; however, Ti peaks were less sharp. In the case of the chit/EE100 sample, the Ti peaks reached a lower intensity compared to the chitosan coating, which could result from a greater thickness of the chit/EE100 coating. The addition of Eudragit E 100 to the chitosan coating reduced the intensity of the O peak. This oxygen decrease is difficult to explain. Presumably, the porous chitosan coatings contain a lot of molecular oxygen, and the addition of Eudragit may be placed inside the empty spaces at the expense of oxygen. The spectra obtained confirmed the absence of other elements in the prepared samples, indicating no contamination during the EPD process.

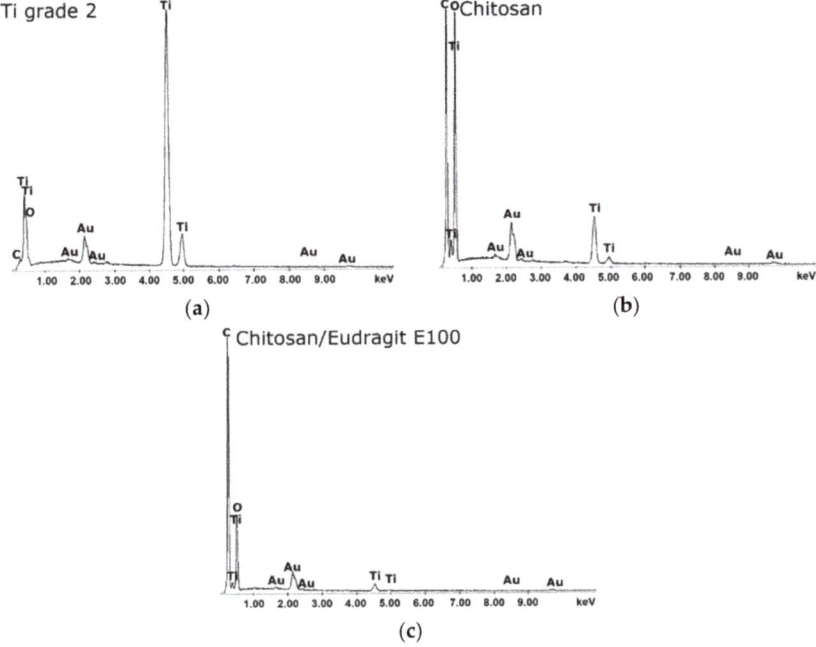

Figure 2. X-ray energy dispersion spectroscopy spectra of the Ti grade 2 substrate (a), the sample with the chitosan coating (b), and (c) sample with chitosan/Eudragit E 100 coating (A1 sample).

Figure 3a depicts the X-ray diffractograms of analyzed specimens. Within the patterns, only peaks associated with the titanium alpha phase can be identified (JCPDS file 44-1294), which indicates the relatively thin, both chitosan and chit/EE100, layers. No reflections of chitosan or Eudragit E 100 can be indexed within the obtained patterns [62,63].

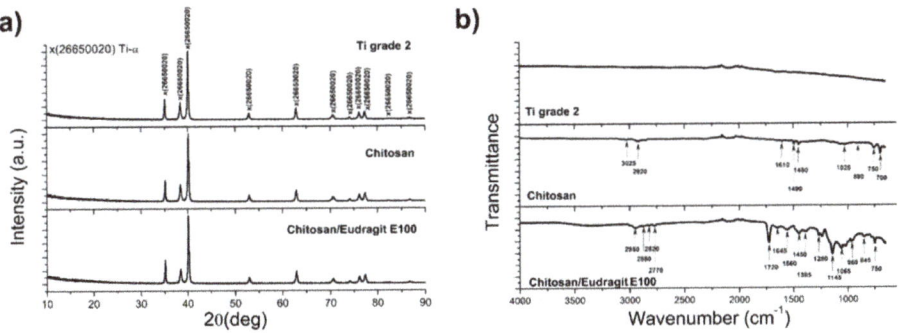

Figure 3. (a) X-ray diffractograms and (b) FTIR spectra of the Ti grade 2 substrate, the chitosan coating, and the chit/EE100 coating (A1 sample).

Figure 3b presents the FTIR results of measured samples. In the case of the spectra acquired for the layered chitosan samples, some low-intensity bands were observed. These bands can be attributed to the chitosan layer. The clear bands which appear in the range of 1680–1480 cm^{-1} can be associated with the vibrations of carbonyl bonds (C=O) of the amide groups, when absorption in the range from 1160 to 1000 cm^{-1} can be recognized as vibrations of CO bonds [64]. In the case of the spectra recorded for the sample with chit/EE100, the bands with higher intensity are visible. FTIR spectrum of Eudragit E 100 presented typical bands of ester groups in the range of 1300–1150 cm^{-1}. A strong C=O ester stretching band was observed at 1720 cm^{-1}. In addition, vibrations of the hydrocarbon chain were observed at 1385, 1450–1490, and 2950 cm^{-1}. Signals visible between 2770 and 2820 cm^{-1} can, on the other hand, be attributed to dimethylethanolamine (DMAE) groups. Such bands were already observed for the stand-alone Eudragit E 100 polymer [65].

3.2. Mechanical Studies

For long-term and load-bearing implants, the mechanical properties are among the most significant factors determining implant durability. The difference between the properties of human bone and the implant can lead to loosening of the implant [66]. Nanoindentation is an increasingly used method for testing thin coatings for biomedical applications [67]. This technique allows for making indents with sizes measured in nanometers, which permits testing thin coatings. It enables the determination of mechanical parameters such as hardness and Young's modulus, and nanoindentation properties: maximum depth of indentation, plastic, and elastic work.

Figure 4 presents single hysteresis load-deformation graphs for the substrate and the prepared coatings. Each of the curves consists of three sections: increasing the force to the maximum value, holding with maximum force, and offloading. A slight deflection on the deformation curves is visible for all tested samples, which results from the temperature drift during the measurement. Based on the obtained curves, nanoindentation parameters were calculated, as presented in Figure 5.

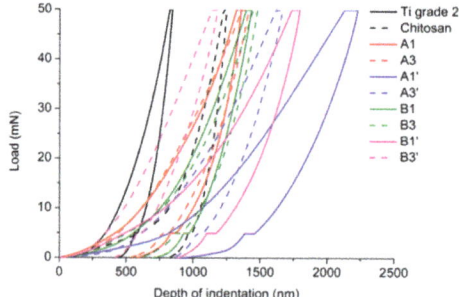

Figure 4. Hysteresis plots of load-deformation for a single indentation measurement for the Ti grade 2 substrate, chitosan, and chit/EE100 coatings.

The Ti grade 2 substrate showed the highest hardness and Young's modulus, which resulted in the smallest indentation depth obtained. All coated samples showed worse mechanical properties, but higher nanoindentation properties as compared to the Ti grade 2 substrate. Similar relationships were observed in the past studies, and they result from the features, like chemical bonds, of the specific material groups. Metals show higher mechanical properties compared to polymers, which leads to a lower depth of indentation [49,68]. In the case of the chitosan coating, the obtained values of hardness and Young's modulus were similar to the values presented in the previous work [49]. Coatings containing EE100 had similar hardness (except A1' and B3') and much lower Young's modulus compared to that of the chitosan coating. This can be explained by the different thickness, packing density, and porosity of chit/EE100 coatings compared to a no-Eudragit coating. The highest hardness of sample B3' results from the application of the highest deposition parameters (time, voltage, EE100 concentration). This coating was probably the thickest and most densely packed. For sample A1′, the lowest hardness value may be due to the thinnest coating and low packing density, due to the short deposition time and lower EE100 concentration in the suspension [69]. The Young's modulus values of chit/EE100 coatings were similar to the value of Young's modulus of human bones. The Young's modulus value closest to Young's modulus of the human tibia cortical bone ($E = 25.8$ GPa) was obtained for sample A1 [70]. This coating showed the highest homogeneity. In implantology, there must be no significant differences in the mechanical properties between the implant and the human bone [66]. The obtained values of parameters determining the mechanical properties of coatings could be influenced by the titanium substrate.

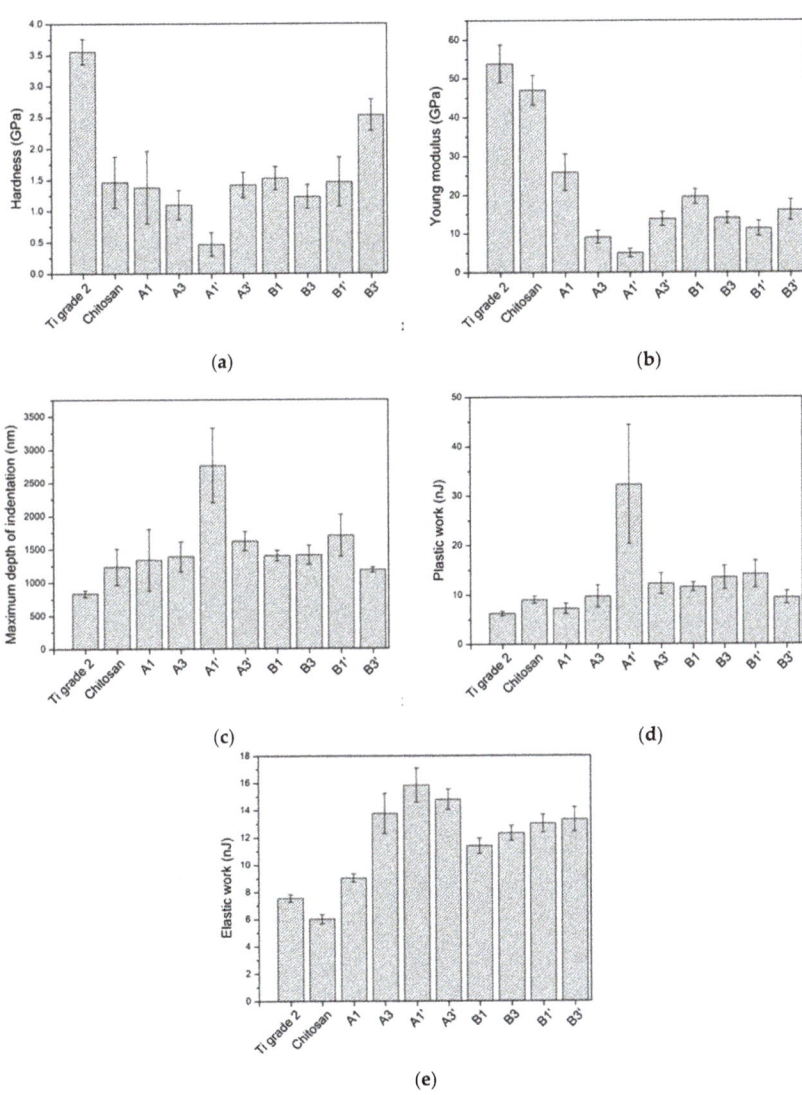

Figure 5. Mechanical properties: (**a**) hardness and (**b**) Young's modulus; nanoindentation properties: (**c**) maximum depth of indentation, (**d**) plastic work and (**e**) elastic work for the Ti grade 2 substrate, the chitosan coating, and the chit/EE100 coatings. Data are presented as the mean ± SD ($n = 10$).

It was difficult to determine the effect of applied coating deposition parameters on the mechanical properties of the prepared coatings. However, for coatings deposited at 10 V, the longer deposition time resulted in a decrease in the hardness and Young's modulus of the coatings. An inverse relationship was observed for 30 V. An increase in the concentration of EE100 in the suspension increased the hardness of the coatings. According to the SEM images (Figure 1), deposition kinetics increased with increased applied voltage, resulting in a more heterogeneous coating structure with visible chit/EE100 clusters that could increase the hardness of coatings locally [71].

The value of elastic work for a particular sample exceeded the value of plastic work. For all samples, the value of plastic work increased with an increased maximum depth of indentation. The obtained results confirm that the tested coatings are more elastic (less brittle) than the reference chitosan coatings, which is a positive impact of Eudragit addition. The increase in plastic work with increasing indent depth results from increasing plastic deformation at the tip, a number of dislocations, and plastic strengthening. High values of standard deviations from the average values probably result from the heterogeneity of the coatings produced as a result of bubble formation during the EPD process [49,58]. There are reports in the literature on a wide range of hardness, Young's modulus, and nanoindentation parameters for chitosan coatings. It results from the differences in applied conditions and measurement parameters [72–74]. Fahim et al. and Akhtar et al. applied much lower loads during the indentation measurements, 3 and 5 mN, respectively, obtaining much lower hardness values of chitosan coatings. Possibly, a too high preliminary load was applied in the case of the conducted tests, and therefore, the results were influenced by the titanium substrate [73,74]. There is no information concerning the mechanical properties of chit/EE100 coatings.

Figure 6 presents plots of the dependence of the friction force on the normal force for each sample with an indication of the critical force causing complete delamination of the coating from the titanium substrate. The value of the critical force was determined based on a comparison of the frictional force dependence on the normal force and optical microscopic observation of the made scratch.

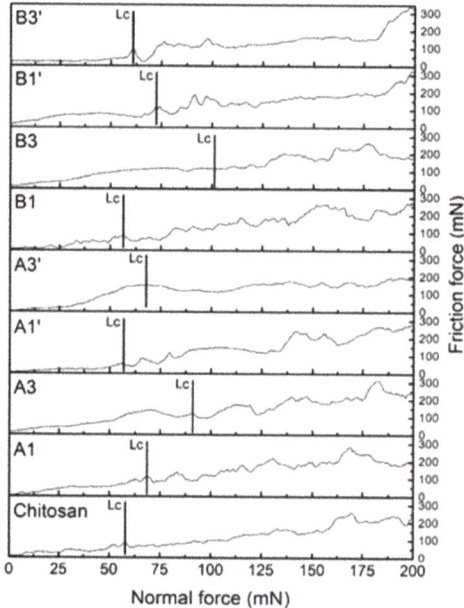

Figure 6. The dependence of the friction force on the normal force obtained for chitosan and chitosan/Eudragit E100 coatings with an indication of the critical force causing complete delamination of the coating from the titanium substrate.

Table 4 shows the values of the average critical load force (Lc) and the corresponding friction force (Lf) determined from scratch test measurements. All coatings with Eudragit E 100 showed higher adhesion to the titanium substrate compared to the chitosan coating. This may be attributed to an increase in the density of chitosan coatings due to the addition of Eudragit E 100 [48]. The most increased adhesion was demonstrated for the A3 and B3 samples. These coatings were prepared at a lower voltage, which resulted in gentle kinetics of biopolymer particle deposition and the formation of a densely packed coating [75]. The effect of the deposition parameters on the critical friction and load is poorly visible. However, there was a tendency (except for samples B1' and B3') that the values of Lc and Lf increased with increasing deposition time. This is probably due to the increase in the thickness of the biopolymer coating. The high values of standard deviations were due to the heterogeneity of the produced coatings. High adhesion of the coating to the implant surface is an important factor during the implant placement procedure, as the implant is exposed to heavy loads that can lead to the removal of the coating [76]. There are almost no studies on the adhesion of chitosan and chit/EE100 coatings to metallic substrates in the literature. Therefore, our data, relating to composite coatings containing chitosan, are difficult to compare [49,77].

Table 4. Nanoscratch test properties of the chitosan and chitosan/Eudragit E 100 coatings (mean ± SD; $n = 10$).

Sample	Nanoscratch Test Properties	
	Critical Load, Lc (mN)	Critical Friction, Lf (mN)
Chitosan	53.87 ± 22.04	61.24 ± 22.04
A1	64.24 ± 25.91	90.73 ± 30.95
A3	91.28 ± 23.06	126.66 ± 46.53
A1'	58.05 ± 8.59	70.28 ± 22.79
A3'	68.18 ± 25.10	105.87 ± 44.77
B1	56.42 ± 23.82	83.34 ± 32.18
B3	90.63 ± 37.58	115.86 ± 48.16
B1'	73.88 ± 15.58	96.55 ± 25.55
B3'	61.00 ± 16.80	84.68 ± 34.24

3.3. Degradation Analysis

The test results of the degradation rate of the investigated coatings are shown in Figure 7. The impact of exposure time and pH on weight loss is visible for both tested samples. The degree of degradation of both the chitosan coating and the chit/EE100 coating increased with the increase in the exposure time and the decrease in the pH of the environment. Similar correlations were reported in other works [78]. Under the influence of lowered pH, the protonation of chitosan and EE100 amine groups intensifies, and as a result of repulsive interaction, the degradation occurs [79,80]. The chit/EE100 coating is significantly more stable at a pH of 7 compared to the coating without Eudragit. The mass loss after 7 days at pH 7 was 1.79% and 32%, respectively. However, the chit/EE100 coating showed greater sensitivity to pH changes. Lowering the pH to 5 caused a sharp increase in the mass loss. The degradation of the coating was comparable at pH 5 and 3. The degradation of the chitosan coating with a decrease in pH was smoother. The chit/EE100 blend contains more amine groups in comparison to chitosan alone. Therefore, the pH reduction results in stronger repulsion of the polymer chains during protonation, resulting in more rapid degradation of the coating [80].

Due to the high stability at neutral pH and high sensitivity to its decline, coatings based on chitosan and Eudragit E 100 could be used in controlled drug delivery systems [81]. The use of this type of biopolymer with, e.g., silver nanoparticles as a coating for implants, would protect against the development of bacterial infection after implantation [82]. Such a system could provide controlled release of the drug substance only at the time of inflammation, which is associated with a decrease in the pH of peri-implant tissues [79]. The high stability of the chit/EE100 coating at neutral pH would also significantly reduce the adverse effect of burst release, i.e., the rapid release of a large dose of the drug after the implant has been placed in an environment simulating body fluids [83].

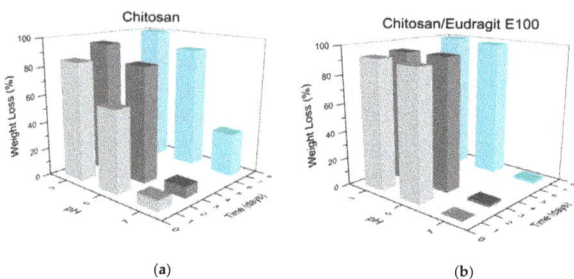

Figure 7. Results of the weight loss (WL) analysis of the (**a**) chitosan and (**b**) chit/EE100 coating (A1 sample).

3.4. Corrosion Studies

Figure 8 depicts potentiodynamic polarization curves obtained for the uncoated Ti grade 2 substrate and chitosan, and chit/EE100 coatings in ASS at 37 °C temperature. Table 5 summarizes the determined corrosion parameters such as open circuit potential, corrosion potential, and current density. Moreover, Figure 9 shows SEM images of the surface topography of the Ti grade 2 substrate, the chitosan coating, and the chit/EE100 coating obtained after corrosion studies. According to the results, samples with coatings showed higher corrosion resistance as measured by corrosion current density compared to the bare Ti grade 2 specimen. Chitosan-based coatings formed a protective layer separating the metallic substrate from the corrosive environment [16]. The addition of EE100 to the chitosan coating slightly reduced its corrosion resistance. This results from a more heterogeneous structure of the chit/EE100 coating and therefore, reduced barrier properties (Figure 9). In the case of samples with coatings, the corrosion potential value was shifted towards positive values compared to the uncoated sample. A slight shift of corrosion potential can be attributed, as confirmed by Tafel curves, to the change of activation polarization, i.e., runs of cathodic and anodic parts. Despite that, the deep decrease in corrosion current can be ascribed mainly to the increasing ohmic resistance of the biopolymer coating as compared to the metallic substrate [55]. Improvement of corrosion resistance of the metallic substrate after application of the chitosan coating was observed in other studies [55,84]. The corrosion resistance of metal implants is crucial because it can affect biocompatibility and mechanical integrity [85]. Implants in aggressive environments are particularly susceptible to corrosion [86]. Titanium is stable in a neutral, alkaline, and only slightly acidic environment; below pH ~ 5, it starts to dissolve. In addition, local pH reduction in peri-implant tissues occurs during inflammation in the human body [87,88]. In such conditions, corrosion products can penetrate peri-implant tissues, which can lead to metallosis and implant rejection [89].

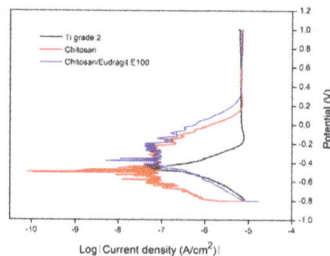

Figure 8. Potentiodynamic polarization curves of uncoated Ti grade 2 substrate, chitosan, and chit/EE100 (A1 sample) coatings in ASS at 37 °C temperature.

Table 5. Open circuit potential, corrosion potential, and current density of the Ti grade 2 substrate and coated substrate with chitosan and chit/EE100 (sample A1).

Sample	OCP (V)	E_{corr} (V)	i_{corr} (nA/cm^2)
Ti grade 2	−0.471	−0.453	794.15
Chitosan	−0.351	−0.445	4.79
Chitosan/EE100 (A1 sample)	−0.306	−0.315	93.79

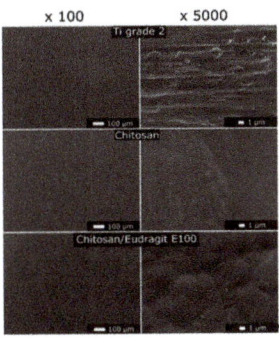

Figure 9. SEM images of the surface topography of the Ti grade 2 substrate, the chitosan coating, and the chit/EE100 coating obtained after corrosion studies; the images obtained at different magnifications, ×100 (on the left) and ×5000 (on the right).

3.5. Contact Angle Studies

Figure 10 shows the values of the average contact angle for the reference Ti grade 2 sample and samples with chitosan and chit/EE100 coatings. The obtained results confirmed the hydrophilic character of all the tested samples. Due to a more uneven surface, almost all samples (except A1′) showed a lower contact angle compared to the reference sample Ti grade 2. The addition of EE100 reduced the wettability of the coating. However, the contact angle was less than 90°. The EE100 coating is water-repellent [62]. The obtained results did not reveal the relationship between the concentration of EE100 in the suspension, the deposition time, or the value of applied voltage and the value of the contact angle. For samples with coatings, a higher surface roughness results in a higher wetting angle. In some cases, surfaces considered to be more uneven were more hydrophilic, probably due to the penetration of water into the irregularities of the coatings [90]. Some studies suggested that for the best cell adhesion, the contact angle of the coatings should be in the range of 40°–60°. However, this range depends on the type of cell and may vary [91]. In the case of bone cells, this range is given as 35°–85°, and the optimum value is 55° [92]. Therefore, all tested samples were close to the upper limit of this requirement.

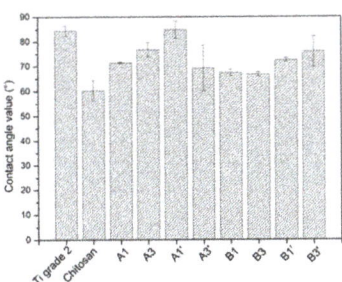

Figure 10. The water contact angle for the Ti grade 2 substrate, the chitosan coating, and the chit/EE100 coatings; data are presented as the mean ± SD (n = 3).

4. Conclusions

The obtained results confirm that it is possible to produce chitosan coatings with the addition of Eudragit E 100 in a one-stage electrophoretic deposition process. Applied deposition parameters affect the quality of the obtained coatings. The increase in the concentration of EE100, voltage, and time of deposition resulted in a more heterogeneous structure of the coatings. The best process deposition parameters for chit/EE100 coating on the surface of Ti grade 2 substrate are the 2.5 g/L of EE100 and 1 g/L of high weight chitosan with the degree of deacetylation > 75 % of 1% (v/v) of the aqueous acetic acid solution, EPD voltage 10 V and EPD time 1 min at room temperature.

Compared to the chitosan coating, the chit/EE100 coatings showed similar hardness and significantly lower Young's modulus, similar to that of a human cortical bone; improved adhesion of the coating to the titanium substrate; a much lower degradation rate at neutral pH. The corrosion resistance and wettability of these coatings were comparable.

The successful deposition of this biopolymer coating, susceptible to pH change, forms a good platform for controlled drug delivery systems, where the antibacterial vector can be silver nanoparticles dispersed in the biopolymer matrix. The production of such a composite coating based on chitosan, EE100, and silver nanoparticles will be the subject of further research. It would be possible to use them as a coating containing a drug substance, e.g., on load-bearing implants, which could limit the adverse effects of the peri-implantitis phenomenon.

Author Contributions: Conceptualization, Ł.P., M.B., and A.Z.; methodology, Ł.P., M.B., A.M.-G., G.S., and M.J.; formal analysis, Ł.P., M.B., A.Z., and A.M.-G.; investigation, Ł.P., M.B., A.M.-G., G.S., and M.J.; writing—original draft preparation, Ł.P., M.B., and A.Z.; writing—review and editing, Ł.P., M.B., and A.Z.; supervision, M.B. and A.Z. All authors have read and agree to the published version of the manuscript.

Funding: This research received no external funding.

Acknowledgments: The authors of the manuscript would like to thank Evonik Industries (Darmstadt, Germany) for the donation of material used for experiments and Robert Kozioł from the Department of Applied Physics and Mathematics, the Gdansk University of Technology for technical assistance in samples preparation.

Conflicts of Interest: The authors declare no conflict of interest.

References

1. Prasad, S.; Ehrensberger, M.; Gibson, M.P.; Kim, H.; Monaco, E.A. Biomaterial properties of titanium in dentistry. *J. Oral Biosci.* **2015**, *57*, 192–199. [CrossRef]
2. Assis, S.L.; Wolynec, S.; Costa, I. The electrochemical behaviour of Ti-13Nb-13Zr alloy in various solutions. *Mater. Corros.* **2008**, *59*, 739–743. [CrossRef]
3. Niemeyer, T.C.; Grandini, C.R.; Pinto, L.M.C.; Angelo, A.C.D.; Schneider, S.G. Corrosion behavior of Ti-13Nb-13Zr alloy used as a biomaterial. *J. Alloys Compd.* **2009**, *476*, 172–175. [CrossRef]
4. Liu, X.; Chu, P.K.; Ding, C. Surface modification of titanium, titanium alloys, and related materials for biomedical applications. *Mater. Sci. Eng. R Rep.* **2004**, *47*, 49–121. [CrossRef]
5. Rau, J.V.; Fosca, M.; Cacciotti, I.; Laureti, S.; Bianco, A.; Teghil, R. Nanostructured Si-substituted hydroxyapatite coatings for biomedical applications. *Thin Solid Films* **2013**, *543*, 167–170. [CrossRef]
6. Souza, J.C.M.; Sordi, M.B.; Kanazawa, M.; Ravindran, S.; Henriques, B.; Silva, F.S.; Aparicio, C.; Cooper, L.F. Nano-scale modification of titanium implant surfaces to enhance osseointegration. *Acta Biomater.* **2019**, *94*, 112–131. [CrossRef] [PubMed]
7. Mistry, S.; Kundu, D.; Datta, S.; Basu, D. Comparison of bioactive glass coated and hydroxyapatite coated titanium dental implants in the human jaw bone. *Aust. Dent. J.* **2011**, *56*, 68–75. [CrossRef]
8. Schmaljohann, D. Thermo- and pH-responsive polymers in drug delivery. *Adv. Drug Deliv. Rev.* **2006**, *58*, 1655–1670. [CrossRef]
9. Sponchioni, M.; Capasso Palmiero, U.; Moscatelli, D. Thermo-responsive polymers: Applications of smart materials in drug delivery and tissue engineering. *Mater. Sci. Eng. C* **2019**, *102*, 589–605. [CrossRef] [PubMed]

10. Wen, J.; Lei, J.; Chen, J.; Gou, J.; Li, Y.; Li, L. An intelligent coating based on pH-sensitive hybrid hydrogel for corrosion protection of mild steel. *Chem. Eng. J.* **2020**, *392*, 123742. [CrossRef]
11. Zhang, A.; Jung, K.; Li, A.; Liu, J.; Boyer, C. Recent advances in stimuli-responsive polymer systems for remotely controlled drug release. *Prog. Polym. Sci.* **2019**, *99*, 101164. [CrossRef]
12. Fu, X.; Hosta-Rigau, L.; Chandrawati, R.; Cui, J. Multi-stimuli-responsive polymer particles, films, and hydrogels for drug delivery. *Chem* **2018**, *4*, 2084–2107. [CrossRef]
13. Kofuji, K.; Qian, C.J.; Nishimura, M.; Sugiyama, I.; Murata, Y.; Kawashima, S. Relationship between physicochemical characteristics and functional properties of chitosan. *Eur. Polym. J.* **2005**, *41*, 2784–2791. [CrossRef]
14. Nikam, V. Eudragit a versatile polymer: A review. *Pharmacologyonline* **2011**, *1*, 152–164.
15. Muxika, A.; Etxabide, A.; Uranga, J.; Guerrero, P.; de la Caba, K. Chitosan as a bioactive polymer: Processing, properties and applications. *Int. J. Biol. Macromol.* **2017**, *105*, 1358–1368. [CrossRef] [PubMed]
16. Simchi, A.; Pishbin, F.; Boccaccini, A.R. Electrophoretic deposition of chitosan. *Mater. Lett.* **2009**, *63*, 2253–2256. [CrossRef]
17. Ahmad, M.; Manzoor, K.; Singh, S.; Ikram, S. Chitosan centered bionanocomposites for medical specialty and curative applications: A review. *Int. J. Pharm.* **2017**, *529*, 200–217. [CrossRef]
18. Ali, A.; Ahmed, S. A review on chitosan and its nanocomposites in drug delivery. *Int. J. Biol. Macromol.* **2018**, *109*, 273–286. [CrossRef]
19. Ahmed, S.; Ikram, S. Chitosan based scaffolds and their applications in wound healing. *Achiev. Life Sci.* **2016**, *10*, 27–37. [CrossRef]
20. Farrokhi-Rad, M.; Shahrabi, T.; Mahmoodi, S.; Khanmohammadi, S. Electrophoretic deposition of hydroxyapatite-chitosan-CNTs nanocomposite coatings. *Ceram. Int.* **2017**, *43*, 4663–4669. [CrossRef]
21. Ordikhani, F.; Simchi, A. Long-term antibiotic delivery by chitosan-based composite coatings with bone regenerative potential. *Appl. Surf. Sci.* **2014**, *317*, 56–66. [CrossRef]
22. Park, J.H.; Saravanakumar, G.; Kim, K.; Kwon, I.C. Targeted delivery of low molecular drugs using chitosan and its derivatives. *Adv. Drug Deliv. Rev.* **2010**, *62*, 28–41. [CrossRef] [PubMed]
23. Jiang, T.; Zhang, Z.; Zhou, Y.; Liu, Y.; Wang, Z.; Tong, H.; Shen, X.; Wang, Y. Surface functionalization of titanium with chitosan/gelatin via electrophoretic deposition: Characterization and cell behavior. *Biomacromolecules* **2010**, *11*, 1254–1260. [CrossRef] [PubMed]
24. Luo, X.L.; Xu, J.J.; Wang, J.L.; Chen, H.Y. Electrochemically deposited nanocomposite of chitosan and carbon nanotubes for biosensor application. *Chem. Commun.* **2005**, *16*, 2169–2171. [CrossRef] [PubMed]
25. Wang, Y.; Guo, X.; Pan, R.; Han, D.; Chen, T.; Geng, Z.; Xiong, Y.; Chen, Y. Electrodeposition of chitosan/gelatin/nanosilver: A new method for constructing biopolymer/nanoparticle composite films with conductivity and antibacterial activity. *Mater. Sci. Eng. C* **2015**, *53*, 222–228. [CrossRef] [PubMed]
26. Franco, P.; de Marco, I. Eudragit: A novel carrier for controlled drug delivery in supercritical antisolvent coprecipitation. *Polymers (Basel)* **2020**, *12*, 234. [CrossRef]
27. Moustafine, R.I.; Kemenova, V.A.; Van den Mooter, G. Characteristics of interpolyelectrolyte complexes of Eudragit E 100 with sodium alginate. *Int. J. Pharm.* **2005**, *294*, 113–120. [CrossRef]
28. Doerdelmann, G.; Kozlova, D.; Epple, M. A pH-sensitive poly(methyl methacrylate) copolymer for efficient drug and gene delivery across the cell membrane. *J. Mater. Chem. B* **2014**, *2*, 7123–7131. [CrossRef]
29. Leopold, C.S.; Eikeler, D. Eudragit® E as coating material for the pH-controlled drug release in the topical treatment of inflammatory bowel disease (IBD). *J. Drug Target.* **1998**, *6*, 85–94. [CrossRef]
30. Farooq, U.; Khan, S.; Nawaz, S.; Ranjha, N.M.; Haider, M.S.; Khan, M.M.; Dar, E.; Nawaz, A. Enhanced gastric retention and drug release via development of novel floating microspheres based on Eudragit E 100 and polycaprolactone: Synthesis and in vitro evaluation. *Des. Monomers Polym.* **2017**, *20*, 419–433. [CrossRef]
31. Świeczko–Żurek, B.; Bartmański, M. Investigations of titanium implants covered with hydroxyapatite layer. *Adv. Mater. Sci.* **2016**, *16*, 78–86. [CrossRef]
32. Cometa, S.; Bonifacio, M.A.; Mattioli-Belmonte, M.; Sabbatini, L.; De Giglio, E. Electrochemical strategies for titanium implant polymeric coatings: The why and how. *Coatings* **2019**, *9*, 268. [CrossRef]
33. Chaurasia, S.; Chaubey, P.; Patel, R.R.; Kumar, N.; Mishra, B. Curcumin-polymeric nanoparticles against colon-26 tumor-bearing mice: Cytotoxicity, pharmacokinetic and anticancer efficacy studies. *Drug Dev. Ind. Pharm.* **2015**, *42*, 694–700. [CrossRef] [PubMed]

34. Selvan, K.; Mohanta, G.; Manna, P.K. Solid-phase preparation and characterization of albendazole solid dispersion. *Ars Pharm.* **2006**, *47*, 91–107.
35. Valizadeh, H.; Zakeri-Milani, P.; Barzegar-Jalali, M.; Mohammadi, G.; Danesh-Bahreini, M.A.; Adibkia, K.; Nokhodchi, A. Preparation and characterization of solid dispersions of piroxicam with hydrophilic carriers. *Drug Dev. Ind. Pharm.* **2007**, *33*, 45–56. [CrossRef]
36. Joshi, G.V.; Kevadiya, B.D.; Bajaj, H.C. Controlled release formulation of ranitidine-containing montmorillonite and Eudragit® E-100. *Drug Dev. Ind. Pharm.* **2010**, *36*, 1046–1053. [CrossRef]
37. Goddeeris, C.; Willems, T.; Houthoofd, K.; Martens, J.A.; Van den Mooter, G. Dissolution enhancement of the anti-HIV drug UC 781 by formulation in a ternary solid dispersion with TPGS 1000 and Eudragit E100. *Eur. J. Pharm. Biopharm.* **2008**, *70*, 861–868. [CrossRef]
38. Elgindy, N.; Samy, W. Evaluation of the mechanical properties and drug release of cross-linked Eudragit films containing metronidazole. *Int. J. Pharm.* **2009**, *376*, 1–6. [CrossRef]
39. Nguyen, C.A.; Konan-kouakou, Y.N.; Allémann, E.; Doelker, E.; Quintanar-guerrero, D.; Fessi, H.; Gurny, R. Preparation of surfactant-free nanoparticles of methacrylic acid copolymers used for film coating. *AAPS PharmSciTech* **2006**, *7*, 63–70. [CrossRef]
40. Lin, S.; Chen, K.; Run-chu, L. Design and evaluation of drug-loaded wound dressing having thermoresponsive, adhesive, absorptive and easy peeling properties. *Biomaterials* **2001**, *22*, 2999–3004. [CrossRef]
41. Prabhushankar, G.L.; Gopalkrishna, B.; Manjunatha, K.M.; Girisha, C.H. Formulation and evaluation of Levofloxacin dental films for periodontitis. *Int. J. Pharm. Pharm. Sci.* **2010**, *2*, 162–168.
42. Basarkar, A.; Singh, J. Poly(lactide-co-glycolide)-polymethacrylate nanoparticles for intramuscular delivery of plasmid encoding interleukin-10 to prevent autoimmune diabetes in mice. *Pharm. Res.* **2009**, *26*, 72–81. [CrossRef] [PubMed]
43. Vibhooti, P.; Rajan, G.; Seema, B. Eudragit and chitosan—The two most promising polymers for colon drug delivery. *Int. J. Pharm. Biol. Arch.* **2013**, *4*, 399–410.
44. Chourasia, M.K.; Jain, S.K. Design and development of multiparticulate system for targeted drug delivery to colon. *Drug Deliv. J. Deliv. Target. Ther. Agents* **2004**, *11*, 201–207. [CrossRef]
45. Moustafine, R.I.; Margulis, E.B.; Sibgatullina, L.F.; Kemenova, V.A.; Van den Mooter, G. Comparative evaluation of interpolyelectrolyte complexes of chitosan with Eudragit® L100 and Eudragit® L100-55 as potential carriers for oral controlled drug delivery. *Eur. J. Pharm. Biopharm.* **2008**, *70*, 215–225. [CrossRef]
46. Xu, B.; Zhang, W.; Chen, Y.; Xu, Y.; Wang, B.; Zong, L. Eudragit® L100-coated mannosylated chitosan nanoparticles for oral protein vaccine delivery. *Int. J. Biol. Macromol.* **2018**, *113*, 534–542. [CrossRef]
47. Ghaffari, A.; Navaee, K.; Oskoui, M.; Bayati, K.; Rafiee-Tehrani, M. Preparation and characterization of free mixed-film of pectin/chitosan/Eudragit® RS intended for sigmoidal drug delivery. *Eur. J. Pharm. Biopharm.* **2007**, *67*, 175–186. [CrossRef]
48. Kouchak, M.; Handali, S.; Naseri Boroujeni, B. Evaluation of the mechanical properties and drug permeability of chitosan/Eudragit RL composite film. *Osong Public Health Res. Perspect.* **2015**, *6*, 14–19. [CrossRef]
49. Bartmański, M.; Pawłowski, Ł.; Zieliński, A.; Mielewczyk-Gryń, A.; Strugała, G.; Cieślik, B. Electrophoretic deposition and characteristics of chitosan/nanosilver composite coatings on the nanotubular TiO_2 layer. *Coatings* **2020**, *10*, 245. [CrossRef]
50. International Standard ISO 4287-1997. *Geometrical Product Specifications (GPS)—Surface Texture: Profile Method – Terms, Definitions and Surface Texture Parameters*; ISO: Geneva, Switzerland, 1997.
51. Loch, J.; Krawiec, H. Corrosion behaviour of cobalt alloys in artificial saliva solution. *Arch. Foundry Eng.* **2013**, *13*, 101–106.
52. Yang, J.; Dahlström, C.; Edlund, H.; Lindman, B.; Norgren, M. pH-responsive cellulose–chitosan nanocomposite films with slow release of chitosan. *Cellulose* **2019**, *26*, 3763–3776. [CrossRef]
53. Lim, H.S.; Hwang, M.J.; Jeong, H.N.; Lee, W.Y.; Song, H.J.; Park, Y.J. Evaluation of surface mechanical properties and grindability of binary Ti alloys containing 5 wt % Al, Cr, Sn, and V. *Metals (Basel)* **2017**, *7*, 487. [CrossRef]
54. Sorkhi, L.; Farrokhi-Rad, M.; Shahrabi, T. Electrophoretic deposition of hydroxyapatite–chitosan–titania on stainless steel 316 L. *Surfaces* **2019**, *2*, 458–467. [CrossRef]
55. Gebhardt, F.; Seuss, S.; Turhan, M.C.; Hornberger, H.; Virtanen, S.; Boccaccini, A.R. Characterization of electrophoretic chitosan coatings on stainless steel. *Mater. Lett.* **2012**, *66*, 302–304. [CrossRef]

56. Sorkhi, L.; Farrokhi-Rad, M.; Shahrabi, T. Electrophoretic deposition of chitosan in different alcohols. *J. Coat. Technol. Res.* **2014**, *11*, 739–746. [CrossRef]
57. Kowalski, P.; Łosiewicz, B.; Goryczka, T. Deposition of chitosan layers on NiTi shape memory alloy. *Arch. Metall. Mater.* **2015**, *60*, 171–176. [CrossRef]
58. Pawlik, A.; Rehman, M.A.U.; Nawaz, Q.; Bastan, F.E.; Sulka, G.D.; Boccaccini, A.R. Fabrication and characterization of electrophoretically deposited chitosan-hydroxyapatite composite coatings on anodic titanium dioxide layers. *Electrochim. Acta* **2019**, *307*, 465–473. [CrossRef]
59. Grandfield, K.; Zhitomirsky, I. Electrophoretic deposition of composite hydroxyapatite-silica-chitosan coatings. *Mater. Charact.* **2008**, *59*, 61–67. [CrossRef]
60. Jugowiec, D.; Kot, M.; Moskalewicz, T. Electrophoretic deposition and characterisation of chitosan coatings on near-β titanium alloy. *Arch. Metall. Mater.* **2016**, *61*, 657–664. [CrossRef]
61. Feng, B.; Weng, J.; Yang, B.C.; Qu, S.X.; Zhang, X.D. Characterization of surface oxide films on titanium and adhesion of osteoblast. *Biomaterials* **2003**, *24*, 4663–4670. [CrossRef]
62. Linares, V.; Yarce, C.J.; Echeverri, J.D.; Galeano, E.; Salamanca, C.H. Relationship between degree of polymeric ionisation and hydrolytic degradation of Eudragit® E polymers under extreme acid conditions. *Polymers (Basel)* **2019**, *11*, 1010. [CrossRef] [PubMed]
63. Abdeen, Z.; Mohammad, S.G.; Mahmoud, M.S. Adsorption of Mn (II) ion on polyvinyl alcohol/chitosan dry blending from aqueous solution. *Environ. Nanotechnol. Monit. Manag.* **2015**, *3*, 1–9. [CrossRef]
64. Dimzon, I.K.D.; Knepper, T.P. Degree of deacetylation of chitosan by infrared spectroscopy and partial least squares. *Int. J. Biol. Macromol.* **2015**, *72*, 939–945. [CrossRef] [PubMed]
65. Kumar, B.P.; Archana, G. Formulation and evaluation of nizatidine solid dispersions. *World J. Pharm. Pharm. Sci.* **2015**, *4*, 810–817.
66. Niinomi, M.; Nakai, M.; Hieda, J. Development of new metallic alloys for biomedical applications. *Acta Biomater.* **2012**, *8*, 3888–3903. [CrossRef]
67. Bartmański, M.; Pawłowski, Ł.; Strugała, G.; Mielewczyk-Gryń, A.; Zieliński, A. Properties of nanohydroxyapatite coatings doped with nanocopper, obtained by electrophoretic deposition on Ti13Zr13Nb alloy. *Materials (Basel)* **2019**, *12*, 3741. [CrossRef]
68. Hryniewicz, T.; Rokosz, K.; Rokicki, R.; Prima, F. Nanoindentation and XPS studies of titanium TNZ alloy after electrochemical polishing in a magnetic field. *Materials (Basel)* **2015**, *8*, 205–215. [CrossRef]
69. Drevet, R.; Jaber, N.B.; Fauré, J.; Tara, A.; Larbi, A.B.C.; Benhayoune, H. Electrophoretic deposition (EPD) of nano-hydroxyapatite coatings with improved mechanical properties on prosthetic Ti6Al4V substrates. *Surf. Coatings Technol.* **2016**, *301*, 94–99. [CrossRef]
70. Sidane, D.; Chicot, D.; Yala, S.; Ziani, S.; Khireddine, H.; Iost, A.; Decoopman, X. Study of the mechanical behavior and corrosion resistance of hydroxyapatite sol-gel thin coatings on 316 L stainless steel pre-coated with titania film. *Thin Solid Films* **2015**, *593*, 71–80. [CrossRef]
71. Wang, Y.C.; Leu, I.C.; Hon, M.H. Kinetics of electrophoretic deposition for nanocrystalline zinc oxide coatings. *J. Am. Ceram. Soc.* **2004**, *87*, 84–88. [CrossRef]
72. Díez-Pascual, A.M.; Gómez-Fatou, M.A.; Ania, F.; Flores, A. Nanoindentation in polymer nanocomposites. *Prog. Mater. Sci.* **2015**, *67*, 1–94. [CrossRef]
73. Fahim, I.S.; Aboulkhair, N.; Everitt, N.M. Nanoindentation investigation on chitosan thin films with different types of nano fillers. *J. Mater. Sci. Res.* **2018**, *7*, 11. [CrossRef]
74. Akhtar, M.A.; Hadzhieva, Z.; Dlouhy, I.; Boccaccini, A.R. Electrophoretic deposition and characterization of functional coatings based on an antibacterial gallium (III)-chitosan complex. *Coatings* **2020**, *10*, 483. [CrossRef]
75. Stevanović, M.; Došić, M.; Janković, A.; Kojić, V.; Vukašinović-Sekulić, M.; Stojanović, J.; Odović, J.; Crevar Sakač, M.; Rhee, K.Y.; Mišković-Stanković, V. Gentamicin-loaded bioactive hydroxyapatite/chitosan composite coating electrodeposited on titanium. *ACS Biomater. Sci. Eng.* **2018**, *4*, 3994–4007. [CrossRef]
76. Brohede, U.; Zhao, S.; Lindberg, F.; Mihranyan, A.; Forsgren, J.; Strømme, M.; Engqvist, H. A novel graded bioactive high adhesion implant coating. *Appl. Surf. Sci.* **2009**, *255*, 7723–7728. [CrossRef]
77. Zhang, J.; Dai, C.S.; Wei, J.; Wen, Z.H. Study on the bonding strength between calcium phosphate/chitosan composite coatings and a Mg alloy substrate. *Appl. Surf. Sci.* **2012**, *261*, 276–286. [CrossRef]
78. Szymańska, E.; Winnicka, K. Stability of chitosan—A challenge for pharmaceutical and biomedical applications. *Mar. Drugs* **2015**, *13*, 1819–1846. [CrossRef]

79. Pawłowski, Ł.; Bartmański, M.; Zieliński, A. pH-dependent composite coatings for controlled drug delivery system—Review. *Inżynieria Mater.* **2019**, *1*, 4–9. [CrossRef]
80. Boeris, V.; Romanini, D.; Farruggia, B.; Picó, G. Interaction and complex formation between catalase and cationic polyelectrolytes: Chitosan and Eudragit E100. *Int. J. Biol. Macromol.* **2009**, *45*, 103–108. [CrossRef]
81. Bagherifard, S. Mediating bone regeneration by means of drug eluting implants: From passive to smart strategies. *Mater. Sci. Eng. C* **2017**, *71*, 1241–1252. [CrossRef]
82. Zheng, K.; Setyawati, M.I.; Leong, D.T.; Xie, J. Antimicrobial silver nanomaterials. *Coord. Chem. Rev.* **2018**, *357*, 1–17. [CrossRef]
83. Thinakaran, S.; Loordhuswamy, A.; Rengaswami, G.V. Electrophoretic deposition of chitosan/nano silver embedded micro sphere on centrifugal spun fibrous matrices—A facile biofilm resistant biocompatible material. *Int. J. Biol. Macromol.* **2020**, *148*, 68–78. [CrossRef] [PubMed]
84. Fayomi, O.S.I.; Akande, I.G.; Popoola, A.P.I. Corrosion protection effect of chitosan on the performance characteristics of A6063 alloy. *J. Bio- Tribo-Corros.* **2018**, *4*, 1–6. [CrossRef]
85. Mareci, D.; Ungureanu, G.; Aelenei, D.M.; Mirza Rosca, J.C. Electrochemical characteristics of titanium based biomaterials in artificial saliva. *Mater. Corros.* **2007**, *58*, 848–856. [CrossRef]
86. Qu, Q.; Wang, L.; Chen, Y.; Li, L.; He, Y.; Ding, Z. Corrosion behavior of titanium in artificial saliva by lactic acid. *Materials (Basel)* **2014**, *7*, 5528–5542. [CrossRef]
87. Chen, Q.; Thouas, G.A. Metallic implant biomaterials. *Mater. Sci. Eng. R Rep.* **2015**, *87*, 1–57. [CrossRef]
88. Surmeneva, M.A.; Sharonova, A.A.; Chernousova, S.; Prymak, O.; Loza, K.; Tkachev, M.S.; Shulepov, I.A.; Epple, M.; Surmenev, R.A. Incorporation of silver nanoparticles into magnetron-sputtered calcium phosphate layers on titanium as an antibacterial coating. *Colloids Surf. B Biointerfaces* **2017**, *156*, 104–113. [CrossRef]
89. Demczuk, A.; Swieczko-Zurek, B.; Ossowska, A. Corrosion resistance examinations of Ti6Al4V alloy with the use of potentiodynamic method in ringer's and artificial saliva solutions. *Adv. Mater. Sci.* **2012**, *11*, 4–11. [CrossRef]
90. Bartmanski, M.; Zielinski, A.; Jazdzewska, M.; Głodowska, J.; Kalka, P. Effects of electrophoretic deposition times and nanotubular oxide surfaces on properties of the nanohydroxyapatite/nanocopper coating on the Ti13Zr13Nb alloy. *Ceram. Int.* **2019**, *45*, 20002–20010. [CrossRef]
91. Heise, S.; Forster, C.; Heer, S.; Qi, H.; Zhou, J.; Virtanen, S.; Lu, T.; Boccaccini, A.R. Electrophoretic deposition of gelatine nanoparticle/chitosan coatings. *Electrochim. Acta* **2019**, *307*, 318–325. [CrossRef]
92. Cordero-Arias, L.; Cabanas-Polo, S.; Gao, H.; Gilabert, J.; Sanchez, E.; Roether, J.A.; Schubert, D.W.; Virtanen, S.; Boccaccini, A.R. Electrophoretic deposition of nanostructured-TiO$_2$/chitosan composite coatings on stainless steel. *RSC Adv.* **2013**, *3*, 11247–11254. [CrossRef]

© 2020 by the authors. Licensee MDPI, Basel, Switzerland. This article is an open access article distributed under the terms and conditions of the Creative Commons Attribution (CC BY) license (http://creativecommons.org/licenses/by/4.0/).

Communication

Degradation Resistance and In Vitro Cytocompatibility of Iron-Containing Coatings Developed on WE43 Magnesium Alloy by Micro-Arc Oxidation

Rongfa Zhang [1],*, Zeyu Zhang [1], Yuanyuan Zhu [1,2], Rongfang Zhao [1], Shufang Zhang [1], Xiaoting Shi [1], Guoqiang Li [1], Zhiyong Chen [1] and Ying Zhao [2],*

[1] School of Materials and Electromechanics, Jiangxi Science and Technology Normal University, Nanchang 330013, China; 18271690030@163.com (Z.Z.); Yuanyuanzhu_01@163.com (Y.Z.); zhaorfamy@126.com (R.Z.); zhang63793@163.com (S.Z.); shixiaoting1001@163.com (X.S.); 18435163505@126.com (G.L.); yong872018744@163.com (Z.C.)
[2] Shenzhen Institutes of Advanced Technology, Chinese Academy of Sciences, Shenzhen 518055, China
* Correspondence: rfzhang-10@163.com (R.Z.); ying.zhao@siat.ac.cn (Y.Z.); Tel./Fax: +86-791-8853-7923 (R.Z.); Tel.: +86-755-8658-5229 (Y.Z.); Fax: +86-755-8658-5222 (Y.Z.)

Received: 6 October 2020; Accepted: 18 November 2020; Published: 23 November 2020

Abstract: Iron (Fe) is an important trace element for life and plays vital functions in maintaining human health. In order to simultaneously endow magnesium alloy with good degradation resistance, improved cytocompatibility, and the proper Fe amount for the body accompanied with degradation of Mg alloy, Fe-containing ceramic coatings were fabricated on WE43 Mg alloy by micro-arc oxidation (MAO) in a nearly neutral pH solution with added 0, 6, 12, and 18 g/L ferric sodium ethylenediaminetetraacetate (NaFeY). The results show that compared with the bare Mg alloy, the MAO samples with developed Fe-containing ceramic coatings significantly improve the degradation resistance and in vitro cytocompatibility. Fe in anodic coatings is mainly present as Fe_2O_3. The increased NaFeY concentration favorably contributes to the enhancement of Fe content but is harmful to the degradation resistance of MAO coatings. Our study reveals that the developed Fe-containing MAO coating on Mg alloy exhibits potential in clinical applications.

Keywords: magnesium alloy; micro-arc oxidation; iron; degradation resistance; cytocompatibility

1. Introduction

Due to the similar specific density and Young's modulus to natural bone, and many Mg ion-associated biological functions in vivo, magnesium alloy is believed to be excellent for biodegradable metallic implants [1–3]. However, too rapid degradation in bilogical environments restricts its clinical applications [1,2]. In fact, the degradation resistance of Mg alloy can be improved by using alloying, surface treatment, or mechanical processing [1,2].

Micro-arc oxidation (MAO), also known as plasma electrolytic oxidation, is an innovative surface treatment method for Mg alloy and titanium alloy [1,4–6]. MAO treatment not only significantly improves the degradation resistance and wear resistance of Mg alloy, but also produces multiple biofunctional ceramic coatings [3]. The properties of MAO coatings are determined by several factors including substrate materials [7] and electrolyte composition and concentration, as well as electrical parameters [3]. Due to the long-term temperature stability, very good mechanical property, and low toxicity of rare earth elements, WE43 alloys are widely used as biomedical Mg alloys [1,8]. During MAO, the electrolyte composition and concentration significantly determine coating properties by affecting the coating chemical composition, phase structure, and surface characteristics (morphology, pore size, and thickness) [1,9]. Besides substrate materials and electrolytes, electrical parameters including frequency, duty cycle, current density, and treatment time are also closely correlated with the coating property [10].

Besides Mg alloy, zinc and iron meet the requirements used for biodegradable orthopedic metals in terms of toxicity, mechanical property, biocompatibility, and corrosion resistance [11]. The degradation products of iron (Fe) mostly consist of Fe oxides and hydroxides, which are biocompatible with the human body [12]. Fe is found in four classes of proteins and enzymes [13]. Fe-containing proteins and enzymes exert a variety of functions including transporting, storing, and activating molecular oxygen [14]. In addition, Fe is an essential element for normal metabolism, growth, development, and maintenance of bones [15,16]. Recent studies show that proper dietary Fe intake may play a positive role in the prevention of osteoporosis in the female subgroup [16], while Fe deficiency adversely affects the cognitive development of children, increases maternal and infant mortality [17], and significantly influences bone mineral density, content, and fragility [18].

Thermal control ceramic coatings containing Fe, Co, Ni, and W elements are black and they are prepared on titanium alloys due to their wide applications in aerospace, satellites, and many other fields [6]. Recently, Fe-containing MAO coatings with improved thermal and optical properties were fabricated by using $Fe_2(SO_4)_3$ on AZ31 Mg alloy [19] and $K_3[Fe(C_2O_4)_3]$ on MB2 Mg alloy [20]. However, the preparation of an Fe-containing coating on medical Mg alloy has not so far been reported. In this study, the highly water-soluble ferric sodium ethylenediaminetetraacetate (abbreviated as NaFeY, Y=[$(OOCCH_2)_2$–N=CH_2CH_2=N$(CH_2COO)_2$]), which is a popular fortificant due to its high bioavailability that can significantly increase Fe and zinc availabilities in porridges [21], was used as the key electrolyte component and Fe-containing biomedical ceramic coatings were firstly fabricated by MAO on Mg alloy. The surface morphology, elemental valence state, and degradation resistance, as well as the in vitro cytocompatibility of MAO-treated samples were measured by scanning electron microscopy (SEM), X-ray photoelectron spectroscopy (XPS), potentiodynamic polarization measurements, and CCK-8 assay.

2. Experimental

Extruded WE43 Mg alloy was provided by Suzhou Chuan Mao Metal Materials Co., LTD (Suzhou, China) and its composition is listed in Table 1.

Table 1. Chemical composition of WE43 Mg alloy (in wt.%).

Element	Y	Zr	Gd	Nd	Cu	Ni	Fe	Mg
Standard value	3.7–4.3	0.4–1.0	0–1.9	2.0–2.5	≤0.02	≤0.005	≤0.01	Balance
Tested value	4.01	0.47	1.72	2.35	0.003	0.004	0.0003	Balance

WE43 Mg alloy was firstly separately machined down to $45 \times 50 \times 8$ mm^3 for surface characteristics and $10 \times 10 \times 8$ mm^3 for electrochemical tests. Prior to MAO treatment, all machined samples were first ground using SiC waterproof abrasive paper from 80 to 3000 grits, then cleaned successively using tap and distilled water. A unipolar constant current mode was used in the study and the applied electrical parameters were a current density of 60 mA/cm^2, a duty cycle of 35%, a pulse frequency of 2000 Hz, and a treating time of 3 min using a homemade MAO5D power supply (Chengdu, China). A sample for MAO treatment and a stainless steel barrel containing the MAO solution were separately connected with the anode and the cathode. The used ammonium bifluoride (NH_4HF_2), hexamethylenetetramine, phosphoric acid (PA), and NaFeY were of analytical reagent grade. Phytic acid (IP6) with a purity of 70% was purchased from Sinopharm Chemical Reagent Co., Ltd. (Shanghai, China). In a nearly neutral base solution containing 6 g/L NH_4HF_2, 360 g/L hexamethylenetetramine, 35 g/L PA, and 8 g/L IP6, samples of 0, 6, 12, and 18 g/L NaFeY were separately added and the fabricated samples were named Fe-0 g/L, Fe-6 g/L, Fe-12 g/L, and Fe-18 g/L, respectively.

After being sputtered with gold, the as-prepared MAO samples were measured by SIGMA scanning electron microscopy (SEM, Zeiss Sigma, Oberkochen, Germany) with an accelerating voltage of 20 kV and coating compositions were analyzed by energy-dispersive spectrometry (EDS) attached to SEM. X-ray photoelectron spectroscopy (XPS, Kratos Analytical, Manchester, UK) was used to determine elemental valence states in MAO coatings. The binding energy values of each element were calibrated according to the adventitious C 1s signal, which was set at 284.6 eV. The degradation resistance was evaluated using a conventional three-electrode electrochemical cell at 37 ± 0.5 °C in Hank's solution, which was composed of 8 g/L NaCl, 0.4 g/L KCl, 0.14 g/L $CaCl_2$, 0.35 g/L $NaHCO_3$, 1.0 g/L $C_6H_{12}O_6$, 0.2 g/L $MgSO_4·7H_2O$, 0.1 g/L KH_2PO_4, and 0.06 g/L $Na_2HPO_4·7H_2O$. At a scanning rate of 1 mV/s using a Gamry Reference 600 electrochemical workstation (Gamry Instruments, Lafayette, IL, USA), potentiodynamic polarization curves were measured from −0.25 V to the open circuit potential (OCP) toward a noble direction until film breakdown.

The cytotoxicity of the as-prepared samples was evaluated by the Cell Counting Kit-8 (CCK-8, Beyotime, Shanghai, China) assay. The mouse pre-osteoblast cells (MC3T3-E1) were seeded on the samples in 96-well culture plates at a density of 5×10^3 per well and incubated for 24 h to allow cell attachment. Then, the culture medium (α-MEM supplemented with 10% FBS) was replaced by extracts (the preparation method according to Reference [22]) supplemented with 10% FBS. After incubating for 1, 3, and 7 days, all samples were transferred to new wells. α-MEM supplemented with 10% CCK-8 was added into each well and cultured for another 2 h. The absorbance was measured by using a microplate reader (Thermo Fisher Scientific, Waltham, MA, USA) at 450 nm. The experiments were independently performed at least in triplicates.

3. Results

The variations in the working voltage with treatment time during the MAO process in the base solution with added 0, 6, 12, and 18 g/L NaFeY are shown in Figure 1.

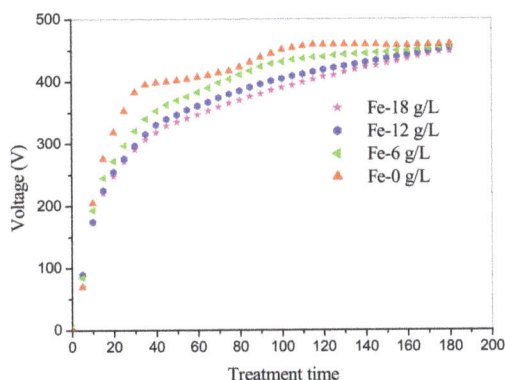

Figure 1. Variations in the working voltage with treatment time in the base solution with added 0, 6, 12, and 18 g/L NaFeY.

In the four solutions containing 0, 6, 12, and 18 g/L NaFeY, the working voltage increased very fast, separately reaching 319, 272, 255, and 248 V during the first 20 s. In the base solution without NaFeY, the voltage went up to 398 V at 40 s and then the increasing rate slowed down. After 145 s, the voltages began to fluctuate. When Mg samples were treated in the solutions containing 6, 12, and 18 g/L NaFeY, the working voltages continually increased. The final voltages in solutions with 0, 6, 12, and 18 g/L NaFeY were 460, 457, 455, and 448 V, respectively (Figure 1). Compared with the base solution without NaFeY, the increased NaFeY concentrations decreased the working voltage but the final voltages did not exhibit evident differences.

Figure 2 shows the surface morphologies and EDS spectra of MAO coatings developed on WE43 alloy. The coating on Fe-0 g/L was uneven and the maximum pore size was about 6.0 μm (Figure 2(a1)). After the addition of NaFeY, it was clear that the pore size and the distance between two adjacent pores decreased (Figure 2(b1–d1)). EDS analysis showed that Fe-0 g/L contained 14.11 at.% C, 51.33 at.% O, 2.43 at.% F, 18.40 at.% Mg, and 13.72 at.% P (Figure 2(a2)). The fabricated Fe-6 g/L, Fe-12 g/L, and Fe-18 g/L were composed of 0.78, 1.53, and 2.27 at.% Fe, respectively (Figure 2(b2–d2)). This suggests that the enhanced Fe content in the coatings is closely pertinent to the increased NaFeY concentration.

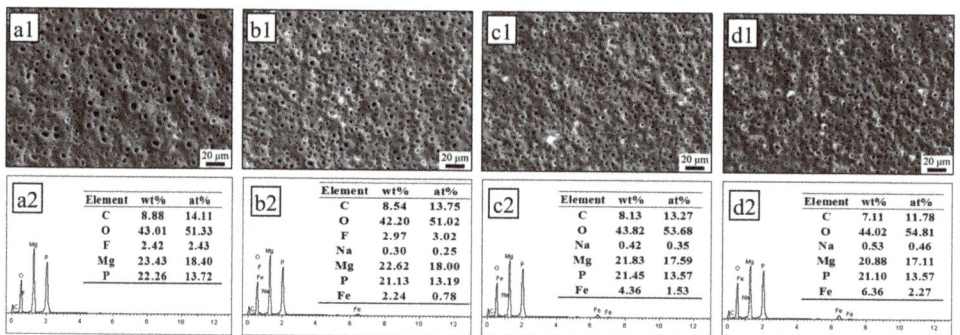

Figure 2. Surface morphologies (**a1–d1**) and chemical compositions (**a2–d2**) of micro-arc oxidation (MAO) coatings fabricated in the base solution with different NaFeY concentrations: (**a1,a2**) Fe-0 g/L; (**b1,b2**) Fe-6 g/L; (**c1,c2**) Fe-12 g/L; (**d1,d2**) Fe-18 g/L.

The high-resolution XPS spectra of O, P, Mg, and Fe elements in Fe-12 g/L are shown in Figure 3. The O 1s spectrum could be divided into three peaks at the binding energy of 530.0, 531.6, and 532.2 eV, corresponding to Fe_2O_3 [14], PO_4^{3-} [23], and OH^- [24], respectively. The binding energy of Mg 1s was centered at 1304.0 and 1304.9 eV (Figure 3b), indicating that the magnesium element in MAO coatings existed as MgO and magnesium phosphate [25], respectively. The P 2p spectrum showed two peaks at 133.4 and 134.1 eV (Figure 3c), assigned to PO_4^{3-} and HPO_4^{2-} [23], respectively. Figure 2d shows that the Fe 2p peaks with a binding energy of $2p_{3/2}$ at 711.1 eV and $2p_{1/2}$ at 724.4 eV correspond to Fe_2O_3 [26].

The potentiodynamic polarization curves of the MAO-treated samples are shown in Figure 4. The relevant electrochemical parameters derived from the potentiodynamic polarization curves are summarized in Table 2. The achieved corrosion current densities(i_{corr}s) of the substrate, Fe-0 g/L, Fe-6 g/L, Fe-12 g/L, and Fe-18 g/L were 1.13×10^{-5}, 6.26×10^{-7}, 8.55×10^{-7}, 9.66×10^{-7}, and 1.24×10^{-6} A/cm^2, respectively, indicating that compared with the substrate, MAO-treated samples significantly improved the degradation resistance. However, with the increase in the NaFeY concentration, the degradation resistance of MAO samples became worse, showing that NaFeY was harmful to the degradation resistance of anodic coatings.

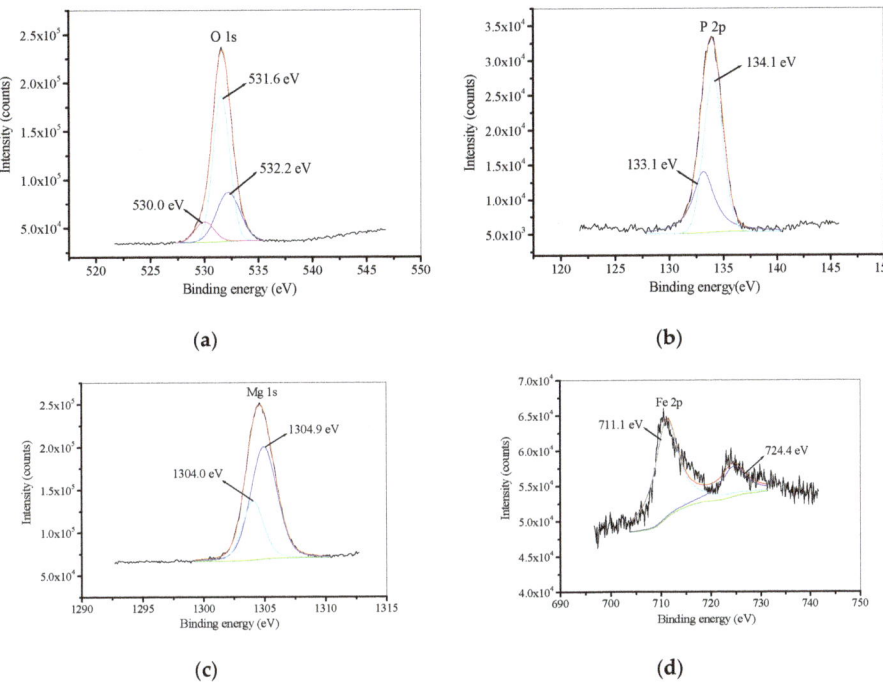

Figure 3. High-resolution XPS spectra of Fe-12 g/L: (**a**) O 1s; (**b**) P 2p; (**c**) Mg 1s; (**d**) Fe 2p.

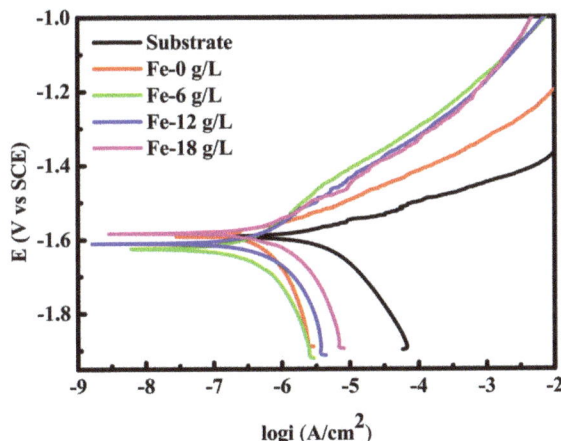

Figure 4. Potentiodynamic polarization curves of MAO samples fabricated in solutions containing different NaFeY concentrations.

Table 2. The electrochemical parameters of MAO samples fabricated in solutions containing different NaFeY concentrations.

Samples	βa (mV/dec)	βc (mV/dec)	i_{corr} (A/cm^2)	E_{corr} (V vs. SCE)
Substrate	313.21	316.26	1.13×10^{-5}	−1.6045
Fe-0 g/L	89.571	383.30	6.26×10^{-7}	−1.5898
Fe-6 g/L	270.08	497.36	8.55×10^{-7}	−1.6260
Fe-12 g/L	220.14	376.65	9.66×10^{-7}	−1.6129
Fe-18 g/L	141.88	311.71	1.24×10^{-6}	−1.5839

The in vitro cytocompatibility of the MAO-treated samples was assessed by CCK-8 assay and the results are shown in Figure 5. Compared with the substrate (Figure 5b), polygonal MC3T3-E1 cells evidently spread on MAO-treated samples with more filose pseudopodium (Figure 5c–f), suggesting that the MAO-treated samples exhibited better initial cell attachment than the substrate. After being cultivated with cells in the extracts for 1, 3, and 7 days, each MAO-treated sample fabricated in four solutions showed significantly higher cell viability than the substrate (Figure 5a). Furthermore, no significant difference could be observed between the MAO-treated samples and the control, suggesting that MAO-treated samples achieved good in vitro cytocompatibility.

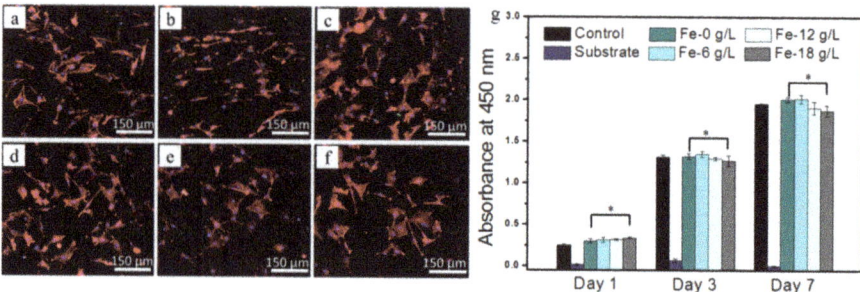

Figure 5. Fluorescent images of MC3T3-E1 cells after culturing for 5 h on (**a**) control, (**b**) substrate, (**c**) Fe-0 g/L, (**d**) Fe-6 g/L, (**e**) Fe-12 g/L, and (**f**) Fe-18 g/L, and (**g**) absorbance values of the MC3T3-E1 pre-osteoblasts cultured for 1, 3, and 7 days on control, substrate, and MAO-treated samples with different Fe amounts. * $p < 0.05$ compared to substrate.

4. Discussion

4.1. Formation of Fe-Containing Ceramic Coating

In order to fabricate an Fe-containing MAO coating on Mg alloy, the selection of the Fe-containing electrolyte is very important. NaFeY, one kind of ferric salt with both excellent solubility in water and chelating ability, was chosen in the study as the Fe-containing electrolyte. In addition, PA and IP6 were used as phosphorus-containing electrolytes. According to Figure 2, the fabricated Fe-0 g/L, Fe-6 g/L, Fe-12 g/L, and Fe-18 g/L were composed of 13.72, 13.19, 13.57, and 13.57 at.% P, respectively (Figure 2(b1–d2)), indicating that both PA and IP6 took part in the coating formation [27,28]. IP6 exhibits a strong chelating ability with positively charged multivalent ions due to its chemical structure, namely six phosphates being connected to one inositol ring [29]. Therefore, in MAO solutions, IP6 competes with FeY^{4-} to combine with Fe^{3+} ions into phytic acid complexes.

According to Figure 3, Mg$_3$(PO$_4$)$_2$ and MgHPO$_4$ were developed in MAO coatings. During MAO, FeY$^-$, PO$_4^{3-}$, and phytic acid complexes migrate toward the anode under the electric field and join in the coating formation according to the following reactions [10,30]:

$$Mg - 2e^- = Mg^{2+} \tag{1}$$

$$3Mg^{2+} + 2PO_4^{3-} = Mg_3(PO_4)_2 \qquad (2)$$

$$Mg^{2+} + HPO_4^{2-} = MgHPO_4 \qquad (3)$$

$$FeY^- \leftrightarrows Fe^{3+} + Y^{4-} \qquad (4)$$

$$Fe^{3+} + 3OH^- = Fe(OH)_3 \qquad (5)$$

$$2Fe(OH)_3 = Fe_2O_3 + 3H_2O \qquad (6)$$

4.2. Degradation Resistance of Fe-Containing Ceramic Coating

According to our previous results, it is difficult to fabricate a uniform MAO coating with a high Fe amount on Mg alloy in an alkaline solution containing NaFeY. Only in solutions containing high concentrations of fluorides and strong alkalines such as NaOH or KOH, can an MAO coating with poor degradation resistance and a low amount of the Fe element be successfully developed on Mg alloy. Fluorides are helpful for coating formation on Mg alloy [30,31], but they are toxic to the environment. In this study, besides a small amount of NH_4HF_2 (6 g/L), 360 g/L hexamethylenetetramine, which has been used as the passive agent of Mg alloy [32], was selected to develop an MAO coating in neutral solutions.

After 6, 12, and 18 g/L NaFeY samples were separately added into the base solution, the i_{corr}s of the fabricated MAO samples continually increased, suggesting that FeY^- was harmful to the degradation resistance of MAO-treated samples. According to Figure 1, NaFeY concentrations did not significantly influence the working voltage, suggesting that NaFeY may influence the coating degradation resistance by surface characteristics. Our recent results show that with the increase in Na_2CaY concentrations, the corrosion resistance of MAO samples becomes worse, attributed to the continually decreased coating thickness [28]. As an organic additive agent with a strong chelating ability, EDTA can improve the uniformity of MAO coatings but decrease the coating thickness [33]. Combined with the previous reports [28,33] and our present results, it can be concluded that Y^{4-} decreases the degradation resistance of MAO-treated Mg alloy. However, the mechanism is not clear and future work will be conducted to clarify it.

4.3. In Vitro Cytocompatibility of Fe-Containing Ceramic Coating

The cytocompatibility of an anodic coating is determined by several factors, for example, degradation resistance and chemical composition. Fluorine (F) is an essential trace element and the proper amount is beneficial to human health. However, when the F amount in an anodic coating is too high (≥19.00 at.%), the fabricated sample exhibits high toxicity [28]. In this study, 6 g/L NH_4HF_2 was added as one kind of passive agent, and the F contents in anodic coatings were lower than 3.02 at.% and revealed good cytocompatibility. In addition to the F element, the cytocompatibility of anodic coatings is closely related to Fe presence. The results show that proper Fe ions are beneficial for cell growth [14]. However, in the study, with the increase in Fe contents in MAO coatings, the cytocompatibility did not significantly improve (Figure 5), which may be attributed to the synergistic effect of the coating composition and degradation resistance. The results show that the degradation resistance is positively related to the coating biocompatibility [11]. Once the NaFeY concentration is increased from 6 to 12 and 18 g/L, Fe amounts in anodic coatings continually increase, enhancing the cytocompatibility. However, the increased NaFeY concentration diminished the coating degradation resistance, which resulted in a slight decrease in cytocompatibility. Consequently, Fe-6 g/L with a lower Fe content but higher degradation resistance resulted in better cytocompatibility, whereas Fe-12 g/L or Fe-18 g/L with a higher Fe content exhibited a slightly lower cytocompatibility.

5. Conclusions

As an essential trace element for life, Fe plays vital functions to maintain human health. It is meaningful to prepare Fe-containing coatings on Mg alloy with improved degradation resistance and in vitro cytocompatibility. In a nearly neutral base solution with 6 g/L NH_4HF_2, 360 g/L hexamethylenetetramine, 35 g/L PA, and 8 g/L IP6, Fe-containing coatings were successfully fabricated on WE43 Mg alloy by MAO treatment. The effects of NaFeY concentrations on the MAO process, surface morphology, chemical composition, elemental valence state, degradation, and in vitro cytocompatibility were investigated. Some conclusions are drawn as follows:

(1) In the base solution with 0, 6, 12, and 18 g/L NaFeY, the final voltages were 460, 457, 455, and 448 V, respectively. The developed MAO coatings in solutions with 6, 12, and 18 g/L NaFeY contained 0.78, 1.53, and 2.27 at.% Fe, respectively. Fe is mainly present as Fe_2O_3 in MAO coatings.

(2) Compared with the bare sample, the developed Fe-containing MAO coatings significantly improve the degradation resistance and in vitro cytocompatibility. The increased NaFeY concentration is favorable to the enhancement of the Fe content but harmful to the degradation resistance of MAO coatings.

(3) The cytocompatibility of MAO-treated samples is synergistically determined by the degradation resistance and chemical compositions. MAO-treated samples with low Fe amounts (Fe-6 g/L) achieved better cytocompatibility than those with higher Fe amounts (Fe-12 g/L or Fe-18 g/L).

Author Contributions: Conceptualization, R.Z. (Rongfa Zhang) and Y.Z. (Ying Zhao); software, G.L.; formal analysis, X.S. and Y.Z. (Yuanyuan Zhu); investigation, Z.Z. and Z.C.; writing—original draft preparation, Z.Z.; writing—review and editing, S.Z. and R.Z. (Rongfang Zhao). All authors have read and agreed to the published version of the manuscript.

Funding: The study was financially supported by National Natural Science Foundation of China (51661010, 81572113, 51861007), National College Students Innovative training program (202011318010) and Scientific Research Fund of Jiangxi Provincial Education Department (GJJ180617).

Acknowledgments: We appreciate the valuable comments provided by other members of our laboratories.

Conflicts of Interest: The authors declare no conflict of interest.

References

1. Narayanan, T.S.; Park, I.S.; Lee, M.H. Strategies to improve the corrosion resistance of microarc oxidation (MAO) coated magnesium alloys for degradable implants: Prospects and challenges. *Prog. Mater. Sci.* **2014**, *60*, 1–71.
2. Yin, Z.Z.; Qi, W.C.; Zeng, R.C.; Chen, X.B.; Gu, C.D.; Guan, S.K.; Zheng, Y.F. Advances in coatings on biodegradable magnesium alloys. *J. Magn. Alloy.* **2020**, *8*, 42–65.
3. Wang, Y.M.; Wang, F.H.; Xu, M.J.; Zhao, B.; Guo, L.X.; Ouyang, J.H. Microstructure and corrosion behavior of coated AZ91 alloy by microarc oxidation for biomedical application. *Appl. Surf. Sci.* **2009**, *255*, 9124–9131.
4. Simchen, F.; Sieber, M.; Kopp, A.; Lampke, T. Introduction to Plasma Electrolytic Oxidation—An Overview of the Process and Applications. *Coatings* **2020**, *10*, 628. [CrossRef]
5. Yao, Z.P.; Jia, F.Z.; Tian, S.J.; Li, C.X.; Jiang, Z.H.; Bai, X.F. Microporous Ni-Doped TiO_2 film Photocatalyst by Plasma Electrolytic Oxidation. *ACS Appl. Mater. Interfaces* **2010**, *9*, 2617–2622.
6. Yao, Z.P.; Hu, B.; Shen, Q.X.; Niu, A.X.; Jiang, Z.H.; Su, P.B.; Ju, P.F. Preparation of black high absorbance and high emissivity thermal control coating on Ti alloy by plasma electrolytic oxidation. *Surf. Coat. Technol.* **2014**, *253*, 166–170. [CrossRef]
7. Shi, Z.M.; Song, G.L.; Atrens, A. Influence of the β phase on the Corrosion Performance of Anodised Coatings on Magnesium-aluminium Alloys. *Corros. Sci.* **2006**, *47*, 2760–2777, reprinted in *Surf. Coat. Technol.* **2006**, *201*, 492.
8. Arrabal, R.; Matykina, E.; Viejo, F.; Skeldon, P.; Thompson, G. Corrosion resistance of WE43 and AZ91D magnesium alloys with phosphate PEO coatings. *Corros. Sci.* **2008**, *50*, 1744–1752. [CrossRef]
9. Toulabifard, A.; Rahmati, M.; Raeissi, K.; Hakimizad, A.; Santamaria, M. The Effect of Electrolytic Solution Composition on the Structure, Corrosion, and Wear Resistance of PEO Coatings on AZ31 Magnesium Alloy. *Coatings* **2020**, *10*, 937. [CrossRef]
10. Zhang, R.; Shan, D.; Chen, R.; Han, E. Effects of electric parameters on properties of anodic coatings formed on magnesium alloys. *Mater. Chem. Phys.* **2008**, *107*, 356–363. [CrossRef]

11. Adhilakshmi, A.; Ravichandran, K.; Narayanan, T.S. Protecting electrochemical degradation of pure iron using zinc phosphate coating for biodegradable implant applications. *New J. Chem.* **2018**, *42*, 18458–18468. [CrossRef]
12. Yang, Y.; Zhou, J.; Detsch, R.; Taccardi, N.; Heise, S.; Virtanen, S.; Boccaccini, A.R. Biodegradable nanostructures: Degradation process and biocompatibility of iron oxide nanostructured arrays. *Mater. Sci. Eng. C* **2018**, *85*, 203–213. [CrossRef]
13. Fraga, C.G. Relevance, essentiality and toxicity of trace elements in human health. *Mol. Asp. Med.* **2005**, *26*, 235–244. [CrossRef]
14. Li, M.; Xu, X.C.; Jia, Z.J.; Shi, Y.Y.; Cheng, Y.; Zheng, Y.F. Rapamycin-loaded nanoporous α-Fe_2O_3 as an endothelial favorable and thromboresistant coating for biodegradable drug-eluting Fe stent applications. *J. Mater. Chem. B* **2017**, *6*, 1182–1194.
15. Yang, J.; Zhang, J.; Ding, C.; Dong, D.; Shang, P. Regulation of Osteoblast Differentiation and Iron Content in MC3T3-E1 Cells by Static Magnetic Field with Different Intensities. *Biol. Trace Element Res.* **2018**, *184*, 214–225. [CrossRef]
16. Xiong, Y.; Wei, J.; Zeng, C.; Yang, T.; Li, H.; Deng, Z.; Zhang, Y.; Ding, X.; Yang, Y.; Lei, G. Association between dietary iron intake and bone mineral density: A cross-sectional study in Chinese population. *Nutr. Diet.* **2016**, *5*, 433–440.
17. Jaramillo, A.; Briones, L.; Andrews, M.; Arredondo, M.; Olivares, M.; Brito, A.; Pizarro, F. Effect of phytic acid, tannic acid and pectin on fasting iron bioavailability both in the presence and absence of calcium. *J. Trace Elements Med. Biol.* **2015**, *30*, 112–117. [CrossRef]
18. Yin, Y.; Li, Y.; Li, Q.; Jia, N.; Liu, A.; Tan, Z.; Wu, Q.; Fan, Z.; Li, T.; Wang, L. Evaluation of the Relationship Between Height and Zinc, Copper, Iron, Calcium, and Magnesium Levels in Healthy Young Children in Beijing, China. *Biol. Trace Element Res.* **2017**, *176*, 244–250.
19. Lu, S.; Qin, W.; Wu, X.; Wang, X.; Zhao, G. Effect of Fe^{3+} ions on the thermal and optical properties of the ceramic coating grown in-situ on AZ31 Mg Alloy. *Mater. Chem. Phys.* **2012**, *135*, 58–62. [CrossRef]
20. Song, Z.K.; Wang, X.D.; Cai, Y.R.; Song, Q.Q. Effect of adding $K_3[Fe(C_2O_4)_3]$ on the characteristics of the magnesium alloy micro-arc oxidation coating. *J. Dispers. Sci. Technol.* **2019**, *41*, 1319–1325. [CrossRef]
21. Kruger, J. Replacing electrolytic iron in a fortification-mix with NaFeEDTA increases both iron and zinc availabilities in traditional African maize porridges. *Food Chem.* **2016**, *205*, 9–13. [CrossRef]
22. Wang, Y.; Lou, J.; Zeng, L.; Xiang, J.; Zhang, S.; Wang, J.; Xiong, F.; Li, C.; Zhao, Y.; Zhang, R. Osteogenic potential of a novel microarc oxidized coating formed on Ti6Al4V alloys. *Appl. Surf. Sci.* **2017**, *412*, 29–36. [CrossRef]
23. Zhu, X.; Chen, J.; Scheideler, L.; Reichl, R.; Geis-Gerstorfer, J. Effects of topography and composition of titanium surface oxides on osteoblast responses. *Biomater.* **2004**, *25*, 4087–4103. [CrossRef]
24. Frateur, I.; Carnot, A.; Zanna, S.; Marcus, P. Role of pH and calcium ions in the adsorption of an alkyl N-aminodimethylphonate on steel: An XPS study. *Appl. Surf. Sci.* **2006**, *252*, 2757–2769.
25. Li, G.Q.; Wang, Y.P.; Zhang, S.F.; Zhao, R.F.; Zhang, R.F.; Li, X.Y.; Chen, C.M. Investigation on entrance mechanism of calcium and magnesium into micro-arc oxidation coatings developed on Ti-6Al-4V alloys. *Surf. Coat. Technol.* **2019**, *378*, 124951.
26. Yamashita, T.; Hayes, P. Analysis of XPS spectra of Fe^{2+} and Fe^{3+} ions in oxide materials. *Appl. Surf. Sci.* **2008**, *254*, 2441–2449. [CrossRef]
27. Zeng, R.C.; Cui, L.Y.; Jiang, K.; Liu, R.; Zhao, B.D.; Zheng, Y.F. In vitro corrosion and cytocompatibility of a micro-arc oxidation coating and poly(L-lactic acid) composite coating on Mg-1Li-1Ca alloy for orghopedic implants. *ACS Appl. Mater. Inter.* **2016**, *8*, 10014–10028.
28. Shi, X.; Wang, Y.; Li, H.; Zhang, S.; Zhao, R.; Li, G.; Zhang, R.; Sheng, Y.; Cao, S.; Zhao, Y.; et al. Corrosion resistance and biocompatibility of calcium-containing coatings developed in near-neutral solutions containing phytic acid and phosphoric acid on AZ31B alloy. *J. Alloy. Compd.* **2020**, *823*, 153721. [CrossRef]
29. Kumar, V.; Sinha, A.K.; Makkar, H.P.; Becker, K. Dietary roles of phytate and phytase in human nutrition: A review. *Food Chem.* **2010**, *120*, 945–959. [CrossRef]
30. Simchen, F.; Sieber, M.; Mehner, T.; Lampke, T. Characterisation Method of the Passivation Mechanisms during the pre-discharge Stage of Plasma Electrolytic Oxidation indicating the Mode of Action of Fluorides in PEO of Magnesium. *Coatings* **2020**, *10*, 965. [CrossRef]

31. Zhu, Y.Y.; Chang, W.H.; Zhang, S.F.; Song, Y.W.; Huang, H.D.; Zhao, R.F.; Li, G.Q.; Zhang, R.F.; Zhang, Y.J. Investigation on Corrosion Resistance and Formation Mechanism of a P–F–Zr Contained Micro-Arc Oxidation Coating on AZ31B Magnesium Alloy Using an Orthogonal Method. *Coatings* **2019**, *9*, 197. [CrossRef]
32. Echeverry-Rendon, M.; Duque, V.; Quintero, D.; Robledo, S.M.; Harmsen, M.C.; Echeverria, F. Improved corrosion resistance of commercially pure magnesium after its modification by plasma electrolytic oxidation with organic additives. *J. Biomater. Appl.* **2018**, *5*, 725–740.
33. Shi, L.L.; Xu, Y.J.; Li, K.; Yao, Z.P.; Wu, S.Q. Effect of additives on structure and corrosion resistance of ceramic coatings on Mg–Li alloy by micro-arc oxidation. *Curr. Appl. Phys.* **2010**, *10*, 719–723. [CrossRef]

Publisher's Note: MDPI stays neutral with regard to jurisdictional claims in published maps and institutional affiliations.

© 2020 by the authors. Licensee MDPI, Basel, Switzerland. This article is an open access article distributed under the terms and conditions of the Creative Commons Attribution (CC BY) license (http://creativecommons.org/licenses/by/4.0/).

Review

Electrodeposited Biocoatings, Their Properties and Fabrication Technologies: A Review

Andrzej Zieliński * and Michał Bartmański

Department of Materials Engineering and Bonding, Gdańsk University of Technology, Narutowicza 11/12, 80-233 Gdańsk, Poland; micbartm@pg.edu.pl
* Correspondence: azielins@pg.edu.pl; Tel.: +48-501-329-368

Received: 27 July 2020; Accepted: 9 August 2020; Published: 12 August 2020

Abstract: Coatings deposited under an electric field are applied for the surface modification of biomaterials. This review is aimed to characterize the state-of-art in this area with an emphasis on the advantages and disadvantages of used methods, process determinants, and properties of coatings. Over 170 articles, published mainly during the last ten years, were chosen, and reviewed as the most representative. The most recent developments of metallic, ceramic, polymer, and composite electrodeposited coatings are described focusing on their microstructure and properties. The direct cathodic electrodeposition, pulse cathodic deposition, electrophoretic deposition, plasma electrochemical oxidation in electrolytes rich in phosphates and calcium ions, electro-spark, and electro-discharge methods are characterized. The effects of electrolyte composition, potential and current, pH, and temperature are discussed. The review demonstrates that the most popular are direct and pulse cathodic electrodeposition and electrophoretic deposition. The research is mainly aimed to introduce new coatings rather than to investigate the effects of process parameters on the properties of deposits. So far tests aim to enhance bioactivity, mechanical strength and adhesion, antibacterial efficiency, and to a lesser extent the corrosion resistance.

Keywords: coatings; electrocathodic deposition; electrophoretic deposition; plasma electrochemical oxidation; electro-spark deposition; electro-discharge deposition; bioactivity; antibacterial efficiency; mechanical strength; corrosion resistance

1. Introduction

The coatings are widely applied for different purposes in the national economies and households. The metallic and polymeric coatings as paints and lacquers are used for the protection of metallic ferrous and nonferrous constructions, in buildings and houses for long term anti-corrosion protection and esthetic effects. The hard coatings are applied to increase the wear resistance. Many functional coatings have been so far proposed and developed. Just recently, for example, gas-barrier thin film coatings for food and beverages [1], the coatings for automotive brake discs [2], tribological coatings [3], functional textile coatings [4], super-hydrophobic anticorrosion coatings [5], and cermet coatings for erosion-corrosion-protection [6] were investigated.

The various coatings have been applied for many years in medicine to improve the properties of the interface between implants and tissues, which determine the biocompatibility and healing time. In past years, the reviews on such functional coatings for dental implantology [7], biocompatible coatings for bone implants [8], ion substituted hydroxyapatite thin films [9], titanium implants polymeric coatings [10], ceramic coatings for osteoporotic bones [11], and pulse laser deposited animal-originated calcium coatings [12] can be found.

The coatings for biological applications, called biocoatings, may be made of metals, polymers, ceramics, and bioglasses, or can be composite coatings, co-deposited, or formed layer-by-layer (hybrid or sandwich coatings). They may be formed by various techniques such as direct electrocathodic deposition (ECD), pulse electrocathodic deposition (PED), electrophoretic deposition (EPD), plasma electrochemical oxidation (PEO) called also micro-arc oxidation (MAO) in calcium, and phosphorus-containing phosphate solutions, chemical vapor deposition (CVD), plasma vapor deposition (PVD), magnetron sputtering, pulsed laser deposition, and many others. They may coat the solid or porous substrates. The biocoatings obtained by electrodeposition are included in many reviews and it is interesting to look for the change of the most popular fabrication techniques from the past till now. Some time ago [13], typical coating methodologies were the ion beam assisted deposition, plasma spray deposition, pulsed laser physical vapor deposition, magnetron sputtering, sol-gel derived coatings, electrodeposition, micro-arc oxidation, and laser deposition. In [14], described methods such as electrodeposition, electrografting, micro-arc deposition, electropolymerization, and electrophoretic deposition of polymers, metals, metal oxides, and ceramics on surfaces of titanium, stainless steels, magnesium alloys, and cobalt alloys were described In another paper [15], the review on surface treatment of titanium alloys focused on anodization mentioning the plasma electrochemical oxidation (PEO) method. In [16], the sol-gel and electrochemical deposition methods were described for many covered metals. The excellent characterization of calcium phosphates, including their coatings, was given in [17]. The achievements in the deposition of Ca-P coatings by sol-gel, thermal spray, magnetron sputtering, electrophoretic deposition, and micro-arc oxidation were recently reviewed in [18]. Liu et al. [19] exhaustively described the PEO technique, plasma spraying, ion implantation, laser surface treatment, sol-gel method, and friction stir processing for biological applications. The review on surface biofunctionalization on implantable metals by Ca-P coatings obtained in various ways, including electrodeposition, was presented in [20]. For popular in recent years Mg alloys applied for biodegradable implants, the cathodic electrodeposition in [21], and the electrophoretic method in [22] was characterized.

This paper is aimed to critically review the most recent, in significant part, achievements in two fields. In the first part of the review, the different coatings deposited by any electrodeposition method (EDM) are described based on their material composition, i.e., metallic, ceramic, polymer, and composite coatings, with attention paid to the effects of component(s) on the coating properties and its application. The further section shows several, most popular deposition technologies applied for biocoatings, and the process parameters, focusing on their effects on properties of coatings. In particular, the deposition techniques here described are those performed in electrolytes and under electric field, limited to ECD, EPD, PEO, electro-spark deposition (ESD), electro-discharge deposition (EDP), and electropolymerization (EP). Thus, the material effects followed by processing effects are subsequently presented together with, at the end of each of the sections, a short summary of the state-of-the-art and prognosis concerning future research.

2. Biocoatings

2.1. Metallic Coatings

The electrodeposited metallic coatings for medical purposes are rare. In [23], the bismuth nanowires were deposited on glassy carbon substrates by the ECD method for electro-reduction of folic acid and its quantitative determination, constituting in such a way a biosensor. In another report [24], a new electrodeposited platinum-iridium coating on platinum, as an alternative to the iridium oxide, was developed for implanted microelectrode arrays being the essential tools in the field of neural engineering. As to increase the corrosion resistance and improve the hydroxyapatite (HAp) deposition, the tantalum was put on the titanium oxide film and then modified with organophosphonic acids [25]. The strontium coating was obtained in [26], demonstrating its positive effect on cell differentiation. In [19], the graphitic carbon nitride nanosheets were used for the self-templated electrodeposition of copper and copper oxide nanostructures for the detection of glucose. In other work [27], the electrodeposition of gold nanowires performed under potentiostatic control, designed as biosensors, was carried out. Finally, the nickel nanoparticles were electrodeposited on TiO_2 nanotubes by either direct current (DC) or cyclic voltammetry (CV) methods to obtain the ferromagnetic coatings [28].

2.2. Ceramic Coatings

2.2.1. Phosphate Coatings

The calcium phosphates (Ca-P), in particular, hydroxyapatite (HAp) and nanohydroxyapatite (nanoHAp), are widely applied in implantology [17] as bone substitutes or bioactive coatings accelerating the primary fixation time of implants. The healing time and the proper adhesion of the coating to the substrate, even under imposed mechanical stresses, at simultaneous lack of adverse effects to the tissues, resulted in an enormous development of HAp bone substitutes and, later on, nanoHAp coatings. The HAp coatings have been developed for the enhancement of bioactivity on titanium implants, and the high number of earlier papers has been focused on the optimization of coatings. Despite that, the characteristic feature of recent research is the use of other than titanium, metallic, and non-metallic surfaces.

In the past, the Ca-P phosphates, mainly stoichiometric or non-stoichiometric hydroxyapatites, were developed as coatings mostly for titanium and its alloys by an application of a direct voltage or current control. For example, in [29] the evolution of the first stages of the crystallization of an electrochemically deposited calcium carbonate was investigated. During the growth of the initial nuclei, the surface of the electrode was covered progressively by the growth of flat multilayers having triangular faces, ranged from one to several molecular layers of calcium carbonate.

The bioactivity is an essential feature of Ca-P coatings. In an early work [30] the electrodeposited calcium phosphate coatings were transformed into apatites during immersion in the simulated body fluid (SBF) at 36.5 °C for 5 days what is still assumed to be a confirmation of the above property.

Recently, the different methods were applied to improve the properties of coatings and their adhesion to solid titanium and its alloys, associated often by pre-treatment of post-treatment. The chemical treatment with H_2O_2 before PED was also often used, as in research [31], in which the porous coating comprising of HAp was formed on the treated titanium surface. The coating was transformed into carbonate- and calcium-deficient HAp layers with a bonelike crystallinity during further immersion in the SBF. To increase adhesion, Ca-P coatings were obtained on pre-calcified titanium, demonstrating a presence of the TiO_x layer with calcium and hydroxyl groups on the surface after this chemical treatment and resulting in a lower water contact angle and lower surface energy than on unmodified titanium surfaces [32]. As concerns the post-treatment, in [33], titanium was firstly subjected to very short electrodeposition and then post-treated in NaOH solution to improve the adhesion. Adhesion was also observed to improve [34] with the increasing electrodeposition time accompanied by a change in the surface morphology from smooth to plate-like, featuring elongated plates, ribbon-like, and finally sharp needle structures. The adhesion of HAp coating on metallic biomaterials is a crucial parameter, dependent on the coating type, deposition method, and substrate [35]. Among four standard techniques of deposition, namely sol-gel, dip coating, electrochemical deposition, and thermal spraying, adhesion of electrodeposited coatings was reported as the lowest. Such a conclusion explains the current trend to optimize this significant feature of biocoatings.

Several studies on Ca-P coatings on titanium biomaterials were performed in the last period with pulse electrodeposition (PED). For titanium irradiated with a high energy electron beam and subject to PED, the lotus flower-like morphology was observed after deposition in SBF in the presence of H_2O_2 [36]. In [37], in the presence of H_2O_2, the coatings composed of HAp and tricalcium phosphate (TCP) were synthesized by PED, revealing the important dissolution-precipitation characteristics and enhanced corrosion protection properties. In another work [38], the application of ultrasonic waves instead of magnetic stirring during PED resulted in the uniform and refined size of plate-like Ca-P crystals.

The HAp and nanoHAp coatings were deposited by EPD on different substrates, for example on NiTi [39], Ti [40–42], oxidized Ti [43–46], and Ag [47]. The Mg-containing hydroxyapatite coatings were also obtained on Ti alloy by MEO with destination for dental materials [48]. For the same purpose, in [49], the HAp nanoparticles were formed on Ti-Nb-Zr alloys by PED followed previous oxidation to nanotubular oxide surface layer. The simultaneous precipitation and electrodeposition of HAp were claimed to be useful for dentistry and orthopedic applications [50].

The HAp coatings can be formed on porous surfaces (scaffolds). After deposition [51] of the Ca-P coatings on the 3D surface of Ti scaffolds made by selective laser melting, the coating morphology was found to depend on the distance from the surface. The top and the bottom surfaces of SLM-Ti scaffolds exhibited continuous and instantaneous nucleation, but the EPD processes at different depth of SLM-Ti scaffolds did not follow the surface processes because of the non-uniform distribution of the potential and the current inside porous structures.

As a treatment positively influencing adhesion, the presence of a nanotubular oxide layer on titanium was extensively applied. For example, the nanotubular surface with the tube length of 560 nm coated with HAp showed higher cell density, higher live cells, and more spreading of MC3T3-E1 cells than that growing on titanium plate surface [52].

The use of HAp-based composite coatings with antibacterial properties is an important research direction. In [53], the calcium phosphate coatings were prepared and then loaded with an antiseptic agent, chlorhexidine digluconate, by a single-step co-deposition. The coating was effective against *Staphylococcus aureus* and *Escherichia. coli* bacteria strains.

Among other metallic biomaterials, the stainless steels were the object of some investigations. In [54], the HAp coating was deposited on 316L stainless steel (SS) in the presence of a significant amount of H_2O_2 by both the direct and pulsed current electrodeposition methods. In [55], the effects of temperature and H_2O_2 content on the morphology, structure, and composition of the coating were confirmed. Surprisingly, in [56], the HAp coatings plated by ECD on 316L steel, but also Ti and Ti6Al4V alloy, increased corrosion rate, likely because of highly porous coating and crevice corrosion. The PED seems a more efficient method as shown in [57], in which the perfect coatings of biocompatible HAp were obtained at various current densities over the SS316 steel. In later research [58], the phosphate coatings, transformed to hydroxyapatite, had thickness from 300 nm up to 2 μm, depending on the deposition mode (continuous or pulsed reverse), and voltage or current parameters, and microstructure of the flake type demonstrating the spontaneous passivation under anodic polarization, and the corrosion potential improved with the increasing presence of hydroxyapatite phase in the coating. An interesting process to form bi-layer (hybrid) coatings [59], the nanoHAp and phosphorus-rich electroless nickel composite coating as an interlayer, was developed before electrodeposition of pure HA coating on stainless steel.

Another group of biometals deposited with HAp was CoCrMo alloys. In [60], the Ca-P coatings were obtained consisting of fine crystallized HAp. On the other hand, for such coatings [61], the coating was weakly hydrophilic, with the phase angle over 85°.

The next group of biometals deposited with HAp coatings was Ni-Ti alloys. In earlier work [62], such coatings were developed to improve the cardiovascular stents with less restenosis than drug-eluting stents. In another study [63], a bipolar pulsed current was used for the electrochemical deposition resulting in the crystalline coating composed of pure HAp nanowalls. Thorough studies of microstructure [64] showed for PED of Ca-P coatings that plate-like and needle-like morphologies were formed in dilute and concentrated solutions, respectively, and HAp appeared at increasing the pulse current density. The coatings obtained in dilute solution showed the best biocompatibility, the highest cell density, and cell proliferation, explained by the stability of the plate-like coating in biological environments. In [65], the surface modification of the NiTi alloy was accomplished before deposition by anodic oxidation and subsequent heat treatment.

Magnesium alloys have become promising materials in the medical field, particularly in tissue engineering applications. In [66], the HAp coating, electrodeposited on AZ91D magnesium alloy and comprising of brushite, lowered the biodegradation rate of Mg alloy in SBF, being the main aim of use of such coatings. The deposited coatings on Mg-1Ca alloy demonstrated improved corrosion resistance in Hank's SBF [67]. In another work [68], three kinds of Ca-P coatings, brushite, HAp, and the most stable and corrosion-resistant fluoridated hydroxyapatite (FHAp) were fabricated by electrodeposition on a biodegradable Mg-Zn alloy. To effectively control the degradation of magnesium alloys, a uniform nano-hydroxyapatite (nanoHAp) coating was applied on AZ31 magnesium alloy using both direct and pulse voltage electrodeposition methods [69]. In similar research on nanoHAp coatings [70] obtained by PED on the porous Mg-2 wt.% Zn scaffold, the corrosion current density reduced from 8 to 0.15 mA/cm^2. The next study on nanoHAp deposition [71] on Mg-2Zn scaffolds performed by PED and post-treatment with alkaline solution demonstrated better biodegradation behavior and biocompatibility, and adherence and proliferation of the MG63 cells when compared to an uncoated alloy. In a more recent report [72], four electrodeposition methods, i.e., the constant potential method, the pulsed potential method, the constant current method, and the current pulsed method, were evaluated. Using the constant current and the current pulsed techniques, smooth uniform coatings were obtained onto AZ31 alloy for both low and high-frequency pulse. Finally, in [73], the Ca-P coatings on Mg implants were subjected to hydrothermal treatment, anodization, and plasma electrolytic oxidation.

The HAp coatings were seldom electrodeposited on ceramics. In [74], HAp coatings, from 1 to 10 μm in thickness, were prepared on human enamel by electrodeposition. They exhibited an acicular morphology and had a tight contact with the substrate.

The HAp coatings were sometimes deposited on polymer surfaces and scaffolds. By the template-assisted PED, the calcium-deficient polycrystalline HAp nanowires and nanotubes were obtained, nucleating preferentially on gold nanoparticles polycarbonate membrane, as nanowires on smaller Au particles and as nanotubes on 400 nm nanoparticles [75]. HAp, carbonate apatite, and dicalcium phosphate dihydrate (brushite) were deposited on carbon/carbon (C/C) composites [76]. A uniform and dense nanoHAp coating with the nanorod-shaped structure were fabricated on carbon nanotubes (CNTs), applied to reinforce the coating [77]. The ultrasonic-assisted electrodeposition was attempted to prepare uniform and smooth hydroxyapatite HAp coating on the surface of the PLA/PVA filaments, which can be potentially applied as biodegradable bone scaffolds [78]. The crystal's growth stages of HAp went through spherical particles, plate-like, and needle-like crystals when electrodeposited from 15 to 60 min.

2.2.2. Substituted Phosphate Coatings

The substituted hydroxyapatites include fluoridated hydroxyapatite (FHAp) and some other inorganic ions substituting calcium ions. The main reason for such design is an expectation of antibacterial effects and better bioactivity. Different substrates and electrodeposition techniques were applied.

As concerns, the titanium alloys, the Sr, Mg, and Zn substituted hydroxyapatite coatings, achieved by PED, exhibited excellent corrosion resistance on Ti6Al4V alloy, in particular at the prolonged pulse off time [79] and strontium-substituted calcium phosphate coatings on Ti6Al4V alloy demonstrated the best osteoblast cells activity and the osteoclast cells proliferation at 5% Sr [80]. In [81], magnesium-doped HAp (MgHAp) was obtained on titanium with a previously formed nanotubular oxide layer, showing only slightly increasing the bond strength. For the strontium/copper substituted hydroxyapatite (SrCuHAp) coatings on titanium, Cu improved its antimicrobial properties, and Sr enhanced the biocompatibility [82]. In other work [83], strontium and manganese co-substituted hydroxyapatite (SrMnHA) crack-free and dense coatings on titanium were prepared by ECD resulting in the decrease in the corrosion current density, strongly hydrophilic surface and better cell morphology, adhesion, spreading, and proliferation, and expression of alkaline phosphatase (ALP) better than on HAp. In similar research [84], the presence of strontium in nanostructured SrHAp coating deposited by ECD was shown again to positively influence the behavior of cells on a titanium surface, i.e., cell viability, adhesion, cell morphology, and the cytoskeletal structure of bone marrow mesenchymal stem cells (MSCs). For the hydroxyapatite doped with silver and manganese (AgMnHAp), deposited on TiO$_2$ nanotubes [85], corrosion resistance decreased by almost two orders of magnitude, and an excellent antimicrobial efficacy was observed with a 100% reduction in viable cells. The lanthanum and copper substituted hydroxyapatite (La/Cu-HAP) coatings made on titanium by ECD demonstrated excellent biocompatibility and bioactivity [86]. The cobalt-substituted Ca-P coatings on Ti22Nb6Zr alloy [87], protected the surface against corrosion in the physiological environment and promoted the development of the cells. The most recently [88], the Ca-P coatings were developed by pulse current deposition onto Ti6Al4V alloy, and after doped with Zn^{2+}, Mg^{2+}, Sr^{2+}, and Ag^+ ions showed better biocompatibility, but also lowered corrosion resistance compared to no-doped Ca-P coating or bare metal.

In the case of steels as substrates, the deposition of (Zn^{2+}, Cu^{2+}, and Ag^+) substituted FHAp coatings were obtained on the 316L steel by ECD to further enhance the antibacterial efficiency of FHAp [5].

For Mg alloys [89], PED in the presence of H$_2$O$_2$ was applied to improve the corrosion resistance and bioactivity of (nano)FHAP coating on Mg-Zn-Ca alloy. A silicon doped calcium phosphate coating was obtained [90] on AZ31 alloy by pulse electrodeposition demonstrating slow degradation rate in SBF, an enhanced corrosion resistance, good cell growth, and enhanced cell proliferation of MG63 osteoblast-like cells, as well as increased activity of ALP. The magnesium-substituted coatings had excellent biocompatibility and no adverse effect. Recently [91], a Si-HAp coating was achieved by PED on an Mg-Zn-Ca alloy to improve biocompatibility, bioactivity, and osteoconductivity.

The substituted HAp coatings were also deposited on polymer substrates. Strontium and magnesium substituted Ca-P coatings were formed on C/C composites by PED, showing the flake-like morphology with a dense and uniform structure that could induce the formation of apatite layers and decrease the corrosion rate of the C/C composites in simulated body fluid [92]. In other research [93], nanocrystalline zinc HAp was prepared by ECD on a titanium-coated silicone provided bioactive properties based on the fibroblast ingrowth and limiting the number of viable *E. coli*.

2.2.3. Oxides' Coatings

The metal oxides for biological applications were seldom deposited. In [94], PED was applied to co-deposit iridium oxide and human plasma proteins. In [95], a novel electrochemical sensor of creatinine was constructed on the glassy carbon electrode. The graphene oxide (GO) and dopamine were transformed to polydopamine and reduced graphene oxide (rGO) film and finally decorated by PED with the nanoparticles Cu by cyclic voltammetry.

2.3. Polymer Coatings

The most applied polymer for electrodeposited biocoatings was chitosan. In [96], the chitosan layer was deposited by ECD on a previously formed porous TiO$_2$ film layer. It is noteworthy that there was an exciting attempt [97] of electrodeposition of bioabsorbable polymer on NP-eluting stent made of stainless steel. Chitosan-mussel protein coatings were proposed to decrease the biodegradation of Mg in body fluids [98]. Photo cross-linking copolymer based on methacrylates deposited by EPD decreased the release of nickel ions from NiTi reducing cytotoxicity inducing by nickel [99]. The polyether ether ketone (PEEK) coatings [100] demonstrated excellent wear resistance, more than two hundred times greater compared with that of the uncoated Ti-13Nb-13Zr alloy.

2.4. Composite Coatings

2.4.1. Ceramics-Ceramics Coatings

The highest group of such coatings is based on hydroxyapatite-phosphate composition with often some ceramics. In [101], a coating composed of HAp and Ca-P phosphate was synthesized over 316L stainless steel through PED showing an increased corrosion resistance and good bioactivity. HAp and HAp/ZrO$_2$ composite coatings were [102] electrodeposited on 316L stainless steel to improve electrochemical behavior and bonding strength. When adding ZrO$_2$ at a concentration of 10 g/L, the corrosion resistance was improved 30 times, and bonding strength between the coating and the substrate increased from 11.6 MPa in pure HAp to 20.8 MPa in composite coatings. Another approach [103] was a deposition of a composite coating comprised of HAp, Ca-P, and multi-walled carbon nanotubes (MWCNTs) on SS316 steel surface by PED showing increased the overall coating modulus of elasticity in the range of 6–10 GPa, a uniform and fast spreading of cells over the coating surfaces and five times higher corrosion resistance.

The HAp-silicate coatings were also developed. In [104], such porous composite coating was obtained demonstrating higher bond strength and better corrosion resistance, and enhanced proliferation of MC3T3-E1 osteoblast cells grown on the HAp/CaSiO$_3$ coating compared to pure HAp. More complex [105] the electroplated zinc-doped hydroxyapatite-silicate coatings presented porous structure, higher corrosion resistance than bare Ti, and improved cell morphology, adhesion, spreading the proliferation of the MC3T3-E1 cells, and expression of ALP alkaline phosphatase comparing to HAp coating.

The nanocomposite coating comprising of calcium titanate, sodium titanate nanotubes, and rutile was prepared on a titanium substrate by a combination of ECD and chemical post-treatment [106] resulting in similar adhesion strength and increasing hardness and corrosion resistance were increased.

The zinc-halloysite nanotubes/(Sr^{2+}, Sm^{2+}) substituted HAp hybrid coating was obtained on Ti6Al4V alloy by the ECD. The enhanced corrosion resistance due to Zn nanotubular layer, significant in-vitro antibacterial efficiency and cell viability studies, and anti-corrosion stability in SBF was observed [107,108].

The coatings composed of HAp and graphene oxide (GO) [109,110] were deposited by ultrasound-assisted PED on the anodized heat-treated surface of the titanium bringing out the highest nano-hardness (3.08 GPa) and Young's modulus (41.26 GPa). Similar hybrid coating composed of the GO inner layer and GO/MgHAp outer layer was prepared on carbon/carbon composites [111] with the bonding strength between the coating and carbon composite over 80% higher than without the oxide.

Carbon nanotubes (CNTs) were also proposed as material for brittle HAp. In earlier research [112], the CNTs reinforced HAp on the Ti substrate demonstrated high nanohardness, enhanced corrosion resistance, and cell viability. In more recent work [113], the addition of CNTs increased nanohardness over 52% and Young's modulus over 41%, and 16% in the adhesion strength thanks to improving integrity, crystallinity, and decreasing Young's modulus mismatch of this coating with titanium substrate. Such coatings were also prepared on AZ31 magnesium alloy by ECD and EPD [114], resulting in decreasing seriously corrosion density from 44.2 to 0.7 µA/cm^2. In [115], different carbon nanostructures, namely exfoliated graphene, RGO, and MWCNTs were co-deposited along with in-situ formed HAp-Ca-P phases. The MWCNTs were deposited together with HAP also by the EPD technique [116,117].

Some oxides were applied for composite HAp coatings. In [118], HAp-ZrO$_2$-TiO$_2$ nanocomposite coatings were synthesized with new morphologies of electrodeposited HAp, such as micro-nanorods. The corrosion current density of coatings was equal to 0.01 µA/cm^2 while the bare substrate was 1.5 µA/cm^2, and porosity was decreased from 46% for HAp coating to 6% for composite coating. A ZnO nanoprism/Zn substituted HAp hybrid layer was prepared on carbon fibers by a two-step electrodeposition method [119].

The HAP was also deposited with borium nitride (BN) by EPD [120] and bioglass and HAp [121].

2.4.2. Ceramics-Polymer Coatings

A novel implant coating material containing graphene oxide and collagen, and HAp was fabricated on Ti16Nb electrooxidized alloy [122] with increased corrosion resistance, superhydrophilic property, antibacterial resistance, and adhesion strength. In other research [123], a novel poly(3,4-ethylene dioxythiophene) based nanocomposite coatings with different contents of FHAp nanoparticles on a Ti-Nb-Zr alloy were developed demonstrating enhanced hardness, a contact angle, and corrosion protection than the pure polymer coatings. The fabricated nanocomposite coating supported the cell adsorption and proliferation of MG-63 cells and showed antibacterial performance. Another research showed [124] for a composite coating of HAp-polypyrrole, synthesized by PED over stainless steel, the increased corrosion resistance by ten-fold as compared to bare polymer coating and supported the growth of MG63 cells. Finally, in [94], a composite coating of iridium oxide and human plasma proteins was obtained by PED.

Some antibacterial nanocomposite coatings were developed [125] as the HAp-Ag-chitosan monolayers or multilayers containing HAp-chitosan and Ag-chitosan layers, with a thickness in the range of 0–20 µm. As assumed, the Ag^+ release rate from the Ag-chitosan layer can be reduced in the layered structure using a HAp-chitosan layer. The multilayer coatings provided corrosion protection of the stainless steel in physiological solutions. Similar biocomposite coating containing chitosan, silver, and hydroxyapatite was developed on anodized titanium substrate by ECD [126]—6-µm thick. The exhibited antibacterial activity against Gram-positive and Gram-negative bacterial strains was due to the synergistic effect of silver and chitosan, with no toxicity concerning MC3T3-E1 cells.

The electrodeposition of a chitosan/layered double MgAl hydroxides was applied to obtain hydro-membranes for protein release triggered by an electrical signal [127]. Chitosan was also co-deposited with HAp, ZNHAp, and MWCNTS [128].

Another approach based on HAp was attempted by the addition of the poly-l-lysine (PLL) + 3,4-dihydroxy benzyl aldehyde (DHBA) and TiO_2 [129] or PEEK [130]. The octadecyltrichlorosilane layer on HAp coatings was shown to be effective in preventing the acid corrosion of dental samples [131].

2.4.3. Metal-Ceramics Coatings

The nickel-MWCNTs coatings were co-electrodeposited [132]. The chemically shortened nanotubes were behaved as inert particles and embedded into the nickel matrix, while the long functionalized nanotubes showed metallic behavior, and during electrodeposition, they were incorporated into the nickel matrix. The hydroxyapatite-zinc coating was attained [133] using ECD possessing excellent adhesion to the substrate. In [134], a two-stage electrochemical synthesis method was used to prepare antimicrobial Ag-HAp composite coatings. In the first stage, a titanium substrate was coated with HAp by ECD, and in the second stage, silver nanoparticles were deposited onto the HAp layer through electrochemical reduction of aqueous Ag^+ to Ag^0. Similar research was described by Bartmanski et al. for nanoAg [135,136] and nanoCu [137]. The coating composed of nanoHAp, MWCNTs, nanoAg, and nanoCu were developed to obtain antibacterial efficiency, high adhesion, and mechanical strength of coatings [138] also with additions of nanoAg and nanoCu. The ultra-fine Pd-Ag-HAp nanoparticles electrodeposited on protruded TiO_2 layer [139] were considered as plausible for dental implants as increasing their biocompatibility.

The influence of the silver nanoparticle (AgNPs) concentration in solution on the electrodeposition of Zn-Cu/AgNP composite coatings was studied by PED [140]. Antimicrobial tests revealed 95%–100% inhibition of bacterial growth after 10 min of contact for the Zn-Cu/AgNP coating with an AgNPs content of 0.28 wt.% against the *E. coli*.

2.4.4. Metal-Polymer Coatings

In [141], composite coatings consisting of polypyrrole (PPy) and Nb_2O_5 nanoparticles were obtained by ECD on 316L stainless steel resulting in enhanced hardness, superior biocompatibility and enhanced corrosion protection as compared to pure PPy coatings. Other research [142] developed the polyethyleneimine film decorated with gold nanoparticles, electrodeposited on glassy carbon electrode, and forming the platform adequate to immobilize proteins for applications as a tool in biotechnology.

2.4.5. Polymer-Polymer Coatings

In [143], the alginate/chitosan layer-by-layer composite coatings were prepared on titanium substrates. No cytotoxicity was observed, and the alginate coating was more beneficial for cells' growth than chitosan coating. More recently [144], bioactive composite coatings based on polypyrrole (PPy)/chitosan were obtained by in situ electrochemical polymerizations with enhanced surface hydrophilicity and enhanced protective performance compared to pure PPy. Recently, partly biodegradable coating composed of chitosan and Eudragit (copolymer of methacrylic acid used for controlling drug release) was developed [145]. The gallium-modified chitosan/poly(acrylic acid) bilayer was proposed for titanium dental implants to prevent their failures [146]. Polymerized microemulsions of PMMA and soy lecithin on 316L stainless steel were also elaborated [147].

2.5. Effects of Component(s) on Properties of Biocoatings

The coatings used or developed for medical purposes can be made of different materials. That is determined by existing specific needs or, more often, by a demand to increase biocompatibility and bioactivity, and implement antibacterial efficiency. Therefore, the coatings are in a great number designed for metallic implants made of stainless steels, titanium, NiTi alloys.

The one-component metallic coatings were proposed for particular applications, mainly for construction of biosensors and, only in single cases, for others such as the creation of ferromagnetic coatings or as additional layers. This main application is related to the dependence of electrochemical potential of metals (half-cells) on the environmental signals like pH, temperature, and oxygen or hydrogen amount. The use of some oxides as single layers was scarce.

The ceramic coatings, in the past, were developed as simple phosphates (Ca-P: triphosphate (TCP), octacalcium phosphate (OCP), hydroxyapatite (HAp), and others). However, such coatings, even cheap and of good bioactivity, were prone to brittle cracking, often weakly adjacent to the base, and did not prevent the development of biofilm. Four solutions were then proposed: to modify the chemical formula of phosphate, to substitute phosphates by other substances, to create the composite HAp-based coatings, or to look for the best deposition technology and process parameters (as discussed in next section).

Among substituted phosphates, the fluorohydroxyapatite in which a fluoro group substituted a hydroxyl group, was often investigated. Besides, the hydrogen atoms were substituted by several metallic elements such as Zn, Sr, Cu, Ag, and Mn, mainly for titanium made implants. The attempts to decrease the solubility rate of biodegradable Mg alloys for implants by the development of SiHAp should also be noted.

Among one-component polymeric coatings, they were used for different purposes: for example, chitosan for increasing the bioactivity or to decrease the Mg degradation, methacrylates for prevention of Ni release from NiTi alloys, and PEEK as an anti-wear coating for biopairs.

The one-component coatings are usually relatively cheap, but their applications are limited as the obtained surface properties are not fully satisfactory. Therefore, the two or three or even four-component coatings were extensively developed.

The mostly applied are coatings composed of two ceramics such as HAp (or, more and more often, nanoHAp), silicates, oxides (mainly reduced graphene oxide rGO), CNTs (mainly MWCNTs), sometimes other substances as BN or halloysite. These coatings often demonstrated enhanced bioactivity and hardness, but their resistance to cracking, adhesion, and antibacterial action could be acknowledged as satisfactory only in part.

The promising direction is developing the metal-polymer coatings, which are usually based on HA with additions of chitosan, collagen, polypyrrole (Py), and others, sometimes implemented also with oxides. Such coatings may be well adjacent to the surface, stronger compared to pure ceramic coatings, and possess better biological properties. This research direction seems not so far developed, but very promising.

Other composite coatings were less investigated. It noteworthy that the metal-ceramics coatings based on nanoHAp implemented with antibacterial nanometals Ag and Cu were used for orthopedic and dental implants. Among the polymer-polymer coatings it is worth noting the appearance of several biodegradable or protective coatings.

Summarizing, there are two most promising development directions of research coatings. The first scientific goal is to create the coating highly bioactive, biocompatible, of high adhesion to both orthopedic and dental implants, but also resistant to corrosion and mechanical stresses, and antibacterial. Such coatings could be based on (substituted or no) nanoHAp and may include several other components: polymer for better crack resistance and additional biological properties (chitosan mainly recommended as a cheap and efficient component), nanoAg (or AgHAp, SrAgHAp, ZnAgHAp, and MnAgHAp). The second objective, less developed even if very promising, is to create the coatings with some specific functions as smart coatings for biosensors and long-term antibacterial protection, thin multifunctional coatings for stents, and wear-resistant coatings. The effect of used material on coating properties and possible applications is then highly significant, and the influence of applied technology, as demonstrated in other sections, is sometimes important and in other cases rather minor.

Figure 1 demonstrates the electrodeposition method made (EDM) biocoatings. They have been divided at first according to whether they are on or multicomponent coatings, and then they have been listed based on the main materials.

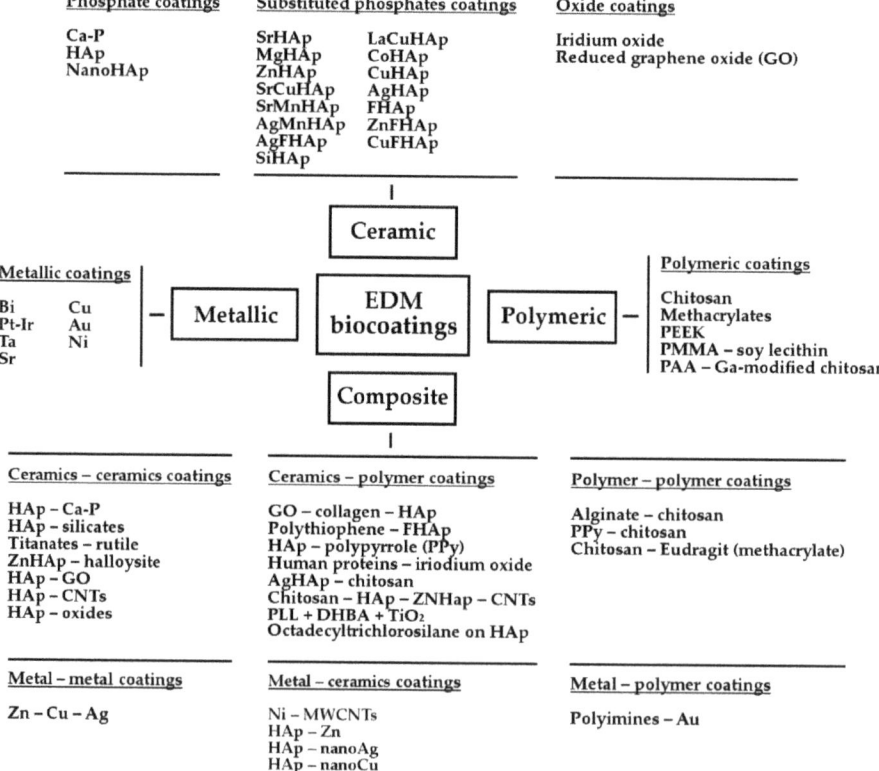

Figure 1. The types of biocoatings created by electrodeposition methods (EDM) as related to the used materials.

3. Deposition Technologies

3.1. Electrocathodic Deposition (ECD)

Among the electrodeposition methods, the electrochemical deposition [16], also called (ECAD) or cathodic deposition, seems the most popular. The crucial determinants of microstructure and properties of coatings include an electrolyte composition, voltage or current, depending on electrodeposition mode, and process time. Some research designs and results are described below. The specific determinants for many investigations are shown in Table 1 for one-component metallic, ceramic and polymer, and composite coatings.

Table 1. The main determinants of the electrocathodic deposition of biocoatings.

Coating; Substrate	Solution and pH	Current Density and/or Voltage Control	Deposition Time and Temperature	Reference
Metallic Coatings				
Bismuth	Bismuth nitrate in acetate pH 4.5	−1.0 V	100 s	[23]
Cu-Cu$_2$O; graphitic carbon nitride	H$_2$SO$_4$, Cu$_2$SO$_4$, Ce and Ni salts	−0.8 V	-	[148]
Ni; Ti	nickel sulfate, nickel chloride, boric acid	50 mA/cm^2	RT 15–60 s	[28]
Ni/Au; PC	NiSO$_4$, H$_3$BO$_3$, KCl (for Ni) Au(CN)$_2$ for Ag	1.4 V	1 min (for Ni)	[27]
Sr; Ti and TiZr	strontium-acetate and acetic acid, Sr-NaF, Sr-NaCl. pH 5	1.3 mA/cm^2	21 °C 60 s	[26]
Pt-Ir	-	-	-	[24]
Ceramic Coatings				
Calcium phosphate (Ca-P); Ti	CaCl$_2$, Ca(H$_2$PO$_4$)$_2$, H$_2$O$_2$. pH 2.5–6.0	−0.8 V vs. SCE.	1.67–60 min 45–65 °C	[149]
Ca-P; Ti	NH$_4$H$_2$PO$_4$, Ca(NO$_3$)$_2$	-	37 °C	[150]
Ca-P; Ti	NH$_4$H$_2$PO$_4$, Ca(NO$_3$)$_2$, NaNO$_3$. pH = 7.4	3.33 mA/cm^2	159 min 24 °C	[106]
Ca-P; Ti	Modified SBF: NaCl, NaHCO$_3$, KH$_2$PO$_4$, pH 7.4	from −1.5 V to −2.5 V vs. SCE	1 h 60 °C	[30,31]
Ca-P; precalcified (NaOH) Ti	Ca(NO$_3$)$_2$, NH$_4$H$_2$PO$_4$. pH = 6	2.5 mA/cm^2	60 min RT	[32]
Ca-P; Ti scaffold	CaCl$_2$, NH$_4$H$_2$PO$_4$	-	-	[51]
Ca-P; Ti-Ni	Three different conditions; (i) Ca(NO$_3$)$_2$, NH$_4$H$_2$PO$_4$, H$_2$O$_2$. pH = 4.3 (ii) and (iii) as above, pH = 6	(i) −0.6 mA/cm^2 (ii) −0.5 mA/cm^2 (iii) −3 mA/cm^2 for 1 s and reverse current 0.1 mA/cm^2 for 2 s	(i) 70 °C (ii) and (iii) 65 °C	[63]
Ca-P; Mg	Ca(NO$_3$)$_2$ KH$_2$PO$_4$ pH = 4.6	3.5 V	90 min 47 °C	[73]
Ca-P; Mg alloy	NaNO$_3$, NH$_4$H$_2$PO$_4$, Ca(NO$_3$)$_2$, H$_2$O$_2$. pH 5	2.5–20 V	40–60 min 20 °C	[72]
Ca-P; Mg-1Ca	Ca(NO$_3$)$_2$, NH$_4$H$_2$PO$_4$. pH = 5	−2.5 V vs. SCE	20–240 min	[67]
HAp; Ti	NH$_4$H$_2$PO$_4$, Ca(NO$_3$)$_2$ pH = 7.2	−2.5 V	10 min 80 °C	[52]
HAp; Ti	Ca(NO$_3$)$_2$, NH$_4$H$_2$PO$_4$.	−1.8 V vs. Ag/AgCl	5 s 80 °C	[33]
HAp; Ti	(i) calcium acetate, acetic acid. (ii) Na$_3$PO$_4$, NaOH pH 9.1	2–4 V	1 h	[151]
HAp	Supersaturated solution of Ca(NO$_3$)$_2$ and NH$_4$H$_2$PO$_4$.	1.5 V	80 °C 1 and 4 h	[50]

Table 1. Cont.

Coating; Substrate	Solution and pH	Current Density and/or Voltage Control	Deposition Time and Temperature	Reference
Ceramic Coatings				
HAp; Ti-6Al-4V	$Ca(NO_3)_2$, $NH_4(H_2PO_4)$, $NaNO_3$. pH 4.2	0.6 mA/cm^2	45 min RT	[75]
HAp; Ti, Ti6Al4V, stainless steels	$Ca(NO_3)_2$, $NH_4(H_2PO_4)$, $NaNO_3$, H_2O_2. pH 5.5	3 V	1 h 85 °C	[56]
HAp; CoCrMo	$Ca(NO_3)_2$, $NH_4H_2PO_4$, H_2O_2 pH = 4.5	3 mA/cm^2	30 min 20 °C	[61]
HAp; CoNiCrMo	$Ca(NO_3)_2$, $NH_4H_2PO_4$.	From -1.4 to -2.2 V versus Ag/AgCl	80 °C	[60]
HAp; Mg	$Ca(NO_3)_2$, $NH_4H_2PO_4$, H_2O_2. pH = 4.3.	4 V	2 h RT	[66]
HAp; Au, PC, stainless steels	$Ca(NO_3)_2$, $NH_4(H_2PO_4)$, H_2O_2. pH 4.5 or 6	120–250 mA/cm^2 or -1.6 V vs. Ag/AgCl	3–10 min 70 °C	[152]
HAp; enamel	$Ca(NO_3)_2$, $NH_4H_2PO_4$, $NaNO_3$	0.5 mA/cm^2	1 h 55 °C	[74]
HAp; PVA/PLA	H_2O_2, $CaCl_2$, KH_2PO_4	2.5–7.5 mA/cm^2	1 h	[78]
FHAp; stainless steels	$CaCl_2$, $NH_4(H_2PO_4)$, NH_4F, H_2O_2 pH = 4.6	1 mA/cm^2	1 min 20–65 °C	[5]
Brushite (DCPD) or FHAp; Mg	$Ca(NO_3)_2$, $NH_4H_2PO_4$. For FHAp, $NaNO_3$ and NaF added. pH = 4.4	5 (DCPD) or 0.5 (FHAp) mA/cm^2	RT (DCPD) or 60 °C (FHAp)	[68]
AgHAp; Ti	NaCl, tris(hydroxymethyl)aminomethane, $CaCl_2$, KH_2PO_4. pH = 7.2	12.5 mA/cm^2	95 °C	[134]
AgMnHAp; Ti	$Ca(NO_3)_2$, $NH_4H_2PO_4$, $Mn(NO_3)_2$, $AgNO_3$	0.09 mA/cm^2	65 °C	[85]
(La/Cu)HAp	$Ca(NO_3)_2$, $La(NO_3)_2$, $Cu(CH_3COO)_2$, $NH_4H_2PO_4$, H_2O_2. pH = 4.5	1.0 mA/cm^2	1 h 65 °C	[86]
MgHAp; nanotubular TiO_2	$Ca(NO_3)_2$, $NH_4H_2PO_4$, $Mg(NO_3)_2$. pH = 4.2	0.85 mA/cm^2	35 min 65 °C	[81]
SrHAp; Ti	$CaCl_2$, $NH_4(H_2PO_4)$, $SrCl_2$, $NaNO_3$	3.0 V	1 h 85 °C	[84]
SrCuHAp	$Ca(NO_3)_2$, $Sr(NO_3)_2$, $CuNO_3)_2$, $NH_4H_2PO_4$. pH = 4.4	0.85 mA/cm^2	30 min 65 °C	[82]
SrMnHAp	$Ca(NO_3)_2$, $Sr(NO_3)_2$, $Mn(NO_3)_2$, $NH_4H_2PO_4$. pH = 4.3	0.85 mA/cm^2	30 min 65 °C	[83]
HAp + CNTs	$Ca(NO_3)_2$, NH_4H_2PO4, H_2O_2 MWCNTs	3 V	pH 4.7	[114]
ZnHAp; Ti	$Ca(NO_3)_2$ NH_4H_2PO $NaNO_3$, H_2O_2.	2.5 V	2 h 85 °C	[133]
ZnHAp; stainless steel	$Ca(NO_3)_2$, $NH_4H_2PO_4$, H_2O_2. pH = 4.5	0.5–3 mA/cm^2	1 h 65 °C	[54]
$CaCO_3$; indium tin oxide	$CaCl_2$, $NaHCO_3$, NaCl. pH = 8.25	-0.86 V	25 °C	[29]

Table 1. Cont.

Coating; Substrate	Solution and pH	Current Density and/or Voltage Control	Deposition Time and Temperature	Reference
Polymer Coatings				
Chitosan; Ti-6Al-4V	CH_3COOH, Chitosan, NaOH pH = 4.75	$0.6\ mA/cm^2$	10 min RT	[96]
Poly (DL-lactide-co-glycolide) (PLGA); stainless steel	PLGA solution	2 mA	-	[97]
Composite Coatings				
Ni-MWCNTs	$NiSO_4$, $NiCl_2$, H_3BO_3, saccharine	$80\ mA/cm^2$	-	[132]
Pd/Ag/HAp	$NH_4H_2PO_4$, NH_4F, HAp, Pd, Ag	23 V	1 h	[139]
NanoHAp-CNTs	CNTs, $NH_4H_2PO_4$, $Ca(NO_3)_2$, $NaNO_3$. pH = 7.4	5 mA	15–30 min 100 °C	[77]
CNTs-HAp	$Ca(NO_3)_2$, K_2HPO_4, CNTs	−1.4 V vs. SCE	1 h	[112]
HAp-CaSiO$_3$	nano-SiO_2, $Ca(NO_3)_2$, $NH_4H_2PO_4$. pH = 4.2	$0.8\ mA/cm^2$	30 min 65 °C	[104]
HAp-CaHPO$_4$; stainless steels	$CaCl_2$; $NH_4H_2PO_4$	5 or 10 mA/cm^2; 1 V, 2 V or 3 V	RT	[58]
ZnHAp-CaSiO$_3$	$Ca(NO_3)_2$, $NH_4H_2PO_4$, $Zn(NO_3)_2$. pH = 4.2	$0.8\ mA/cm^2$	30 min 65 °C	[105]
ZnO/ZnHAp hybrid coating; carbon fiber	$Zn(NO_3)_2$, $Ca(NO_3)_2$, $NH_4H_2PO_4$.	1st stage: $0.6\ mA/cm^2$ (ECD). 2nd stage: 3 V (EPD).	1st stage: 30 min, 343 K 2nd stage: 60 min	[119]
Zn-halloysite nanotubes (HNT)/SrSmHAp hybride coating; Ti6Al4V	$Ca(NO_3)_2$, $Sr(NO_3)_2$, $Sm(NO_3)_2$, $NH_4H_2PO_4$	$1.0\ mA/cm^2$	30 min RT	[108]
Halloysite nanotubes (HNT)-CeHAp; Ti alloy	$Ca(NO_3)_2$, $NH_4H_2PO_4$, $Ce(NO_3)_2$, halloysite nanoclay, HCl. pH = 4.5	-	-	[107]
HAp-Ag-chitosan; Pt, graphite or stainless steel	chitosan solutions containing HAp nanoparticles and dissolved $AgNO_3$	$0.1\ mA/cm^2$	-	[125]
HAp-ZrO$_2$; Ti	$Ca(NO_3)_2$, $NH_4H_2PO_4$, $NaNO_3$, H_2O_2, ZrO_2 particles. pH = 4.5	$1\ mA/cm^2$	45 min 65 °C	[102]
HAp-ZrO$_2$-TiO$_2$	$Ca(NO_3)_2$, $NH_4H_2PO_4$, $NaNO_3$, ZrO_2, TiO_2. pH = 4.2	Constant direct current	85 °C 2 h	[118]
HAp-GO-collagen; Ti-Nb	$Ca(NO_3)_2$, $NH_4H_2PO_4$, GO, collagen in SBF. pH = 4.1–4.3	$2\ mA/cm^2$	60 min 33 °C	[122]
GO (an inner layer)/GO-MgHAp (an outer layer); C/C composites	1st stage (GO inner-layer): Go water suspension. 2nd stage (GO-MGHAp): $NH_4H_2PO_4$, $Ca(NO_3)_2$, $Mg(NO_3)_2$, GO.	1st stage: 30–70 V (EPD). 2nd stage: 3 mA (ECD).	1st stage: 1–7 min (EPD). 2nd stage: 1 h, 50 °C (ECD).	[111]
Chitosan-AgHAp on nanotubular TiO$_2$	$Ca(NO_3)_2$, $NH_4H_2PO_4$, $AgNO_3$.	$0.85\ mA/cm^2$ 35 min at 50 °C	35 min 50 °C	[126]
Polietyleneimine (PEI)-Ag	Hydrogen tetrachlorate (III).	−1.2 V vs. (Ag/AgCl)	45 s	[142]
Polypyrrole-chitosan; stainless steel	Pyrrole in oxalic acid, with and without the addition of chitosan.	15 mA	1 h	[144]
Polyacrylic acid (PAA) followed by Ga-modified chitosan; Ti	PAA water solution	For Ga-modified chitosan: 1.5 V	15–60 min	[146]
Alginate/chitosan (layer-by-layer coating)	Chitosan water solution Alginate dissolved in acetic acid	20 V	20 min	[143]
Chitosan-protein; Mg	Chitosan in acetic acid Proteins in citric acid	$1\ mA/cm^2$	10 min	[98]

RT—room temperature. Empty spaces mean that no data have been reported in references.

3.1.1. Effect of an Electrolyte Composition

As shown in Table 1, the regular baths for the deposition of Ca-P compounds contained such salts as $Ca(NO_3)_2$, $CaCl_2$, $CaSiO_3$, and $(NH_4)H_2PO_4$, often H_2O_2, more seldom KH_2PO_4. The typical concentration of calcium salt was 0.042 mol/L and 0.025 mol/L of $(NH_4)H_2PO_4$. Hydrogen peroxide was added in concentration from 2000 ppm to even 2% [61]. As concerns the other additives, in [114] the addition of 1% MWCNTs to HAp showed the most positive effect on corrosion resistance.

3.1.2. Effect of Deposition Potential

In deposition performed under constant voltage, its value was usually between 1 and 3 V. The impact of potential was investigated in a few works. In [58], at 1 V of voltage, 300 to 400 nm thick flake type of deposition with random orientation was observed together with flakes either parallel or perpendicular to the coating surfaces, and under 2 and 3 V potential, an average thickness of flakes increased to around 2 µm. In [33,60], calcium phosphate coatings on the CoNiCrMo substrate were prepared by electrodeposition at different voltages and the moderate potential sample (−1.8 V vs. AgCl) they exhibited the most uniform coating. In [30], the calcium phosphates were electrodeposited on titanium and the coating electrodeposited at −2.0 V (SCE; Saturated Calomel Electrode) in the modified SBF containing CO_3^{2-} ions was the most bioactive, showing transformation into carbonate apatite similar to bone apatite. As stated in [152], the desired potential range for HAp electrodeposition may be comprised between −1.5 and −2.2 V (Ag/AgCl). These results are like each other and the effect of potential seems moderate.

3.1.3. Effect of Deposition Current

Such studies were carried out usually at current densities beginning well below 1 mA/cm^2 up to even 10 mA/cm^2 or more, as shown in Table 1. In [58], under constant cathode current density of 5 mA/cm^2, the broom-like shape of deposition structure was observed, with flakes precisely perpendicular to the coating surfaces, with one after another originated from a single point with a wall thickness of around 400–500 nm, and spherical-shaped porous apatite structure with a diameter around 800–900 nm along with 50–100 nm pores throughout the outer peripheral surface appeared. At 10 mA/cm^2 current density, the continuous coating, and random flake type deposits were observed with around 1–2 µm wall thickness. In the ultrasonic-assisted electrodeposition of hydroxyapatite [78], when the current density increased from 2.5 to 7.5 mA/cm^2, the grain size decreased from 414 to 265 nm. With the current density increasing from 2.5 to 5.0 mA/cm^2, the Ca/P ratio rises from 1.53 to 1.58. Conversely, the Ca/P ratio drastically decreases to 1.25 at a current density of 7.5 mA/cm^2.

3.1.4. Effect of Deposition Time

The result of deposition time was not often investigated, and applied values were diverse. In [34], by increasing the electrodeposition time from 1 to 30 min, the coating thickness increased, but also the surface morphology of the Ca-P coatings was greatly affected going from smooth to plate-like, featuring elongated plates, ribbon-like to finally sharp needle structures. The optimum electrodeposition procedures leading to both good cell-material interaction and sufficient mechanical properties could be achieved with relatively thin coatings produced at short electrodeposition times. The loading time increased the HAp crystallinity [149].

3.1.5. Effect of Deposition pH

The pH value, a rule, was buffered between 4 and 6. The HAp crystallization process was favored by an increase in pH from 2.5 to 6 [149].

3.1.6. Effect of Deposition Temperature

The deposition processes were conducted either at room temperature or, seldom, about 60–70 °C. In [149], with increasing temperature, the deposited hydroxyapatite was occasionally of plate-like shape, and the width and the length of the deposited calcium phosphates at 65 °C were more extensive than those at 55 °C.

3.2. Pulse Electrocathodic Deposition (PED)

The important determinants are shown in Table 2 for one-component metallic, ceramic and polymer, and composite coatings.

Table 2. The main determinants of the pulse electrocathodic deposition of biocoatings.

Coating; Substrate	Solution	Voltage/Current for Pulsed Mode Deposition	Deposition Time and Temperature	Reference
Metallic Coatings				
Ni; Ti	$NiSO_4$, H_3BO_3. pH 2 or 5	From 0 to -1.5 V at the scan rates of 20 and 50 mV/s	-	[28]
Tantalum	LiF, TaF_5	-2.6 V to 1.6 V	30 s to 2 h	[25]
Ceramic Coatings				
Ca-P; Ti	$Ca(NO_3)_2$, $NH_4H_2PO_4$, H_2O_2, GO. pH 6	Pulsed mode 15 mA/cm^2 duty cycle 0.1	65 °C	[38]
Ca-P; Ti	$Ca(NO_3)_2$ $NH_4H_2PO_4$, H_2O_2. pH 4.3	Pulse mode -1.4 V Duty cycle 0.5	1–30 min	[34]
Ca-P; Ti	$Ca(NO_3)_2$, $NH_4H_2PO_4$, chlorhexidine digluconate. pH 4.2	Pulse mode 2–5 mA/cm^2	40–60 °C 30 min	[53]
Ca-P; Mg alloy	$NaNO_3$, $NH_4H_2PO_4$, $Ca(NO_3)_2$, H_2O_2. pH 5.0	Pulse mode (i) Constant voltage 2.5–20 V or (ii) Constant current density 10–200 mA/cm^2 Duty cycle 0.25–0.75	20 °C (i) 40–60 min (ii) 2–6 h	[72]
Ca-P; NiTi	$Ca(NO_3)_2$, $NH_4H_2PO_4$, H_2O_2. pH 4.3	5, 10, 15, and 20 mA/cm^2 Duty cycle 0.1	25 min	[65]
Ca-P; Ti-Ni	Three different conditions; (i) $Ca(NO_3)_2$, $NH_4H_2PO_4$, H_2O_2. pH 4.3 (ii) and (iii) as above, pH 6	Constant mode (i) 0.6 mA/cm^2, or (ii) -0.5 mA/cm^2 (iii) pulsed mode-3 mA/cm^2 duty cycle 0.33	(i) 70 °C (ii) and (iii) 65 °C	[62]
Ca-P and HAp; stainless steel	$CaCl_2$, $NH_4H_2PO_4$, NaCl.	Pulsed mode 5, 10 and 20 mA/cm^2 duty cycle 0.2	1 h room temperature (RT)	[57]
Polymorphic apatites; C/C composites	$Ca(NO_3)_2$, $NH_4H_2PO_4$.	Pulse mode 3, 5 and 10 V Duty cycle 0.2	60 °C 3 h	[76]
HAp; Ti	$CaCl_2$, K_2HPO_4, H_2O_2.	1 mA/cm^2 duty cycle 0.2 and 0.8	1 h	[36]
HAp; stainless steel	$Ca(NO_3)_2$, $NH_4H_2PO_4$, $NaNO_3$. pH = 5.77	Pulse mode -1.6 V/SCE, scanning rate 5 mV/s	25 °C 26.667 min	[55]
HAp; NiTi	$Ca(NO_3)_2$, $NH_4H_2PO_4$, $NaNO_3$, H_2O_2. pH 6.0	Pulsed mode 3.0 mA/cm^2	25 min 65 °C	[63]
HAp; NiTi	$Ca(NO_3)_2$, $NH_4H_2PO_4$, $NaNO_3$, H_2O_2. pH 4.3	Pulse mode 1.5–15 mA/cm^2 duty cycle 0.2	25 min 70 °C	[64]
HAp; Mg alloys	$Ca(NO_3)_2$, $NH_4H_2PO_4$, Na_2SiO_3, $NaNO_3$. pH 4, 5 or 6	Pulsed mode 40 and 60 mA/cm^2 duty cycle 0.1, 0.2	30 min 25–100 °C	[91]

Table 2. Cont.

Coating; Substrate	Solution	Voltage/Current for Pulsed Mode Deposition	Deposition Time and Temperature	Reference
Ceramic Coatings				
NanoHAp; Mg alloy	$Ca(NO_3)_2$, $NH_4H_2PO_4$, H_2O_2. pH = 4.5	Pulse mode −3V Duty cycle 0.2	RT	[69]
NanoHAp; Mg-Zn scaffold	$Ca(NO_3)_2$, $NH_4H_2PO_4$, $NaNO_3$	20–40 mA/cm^2 Duty cycle 0.1 and 0.2 temperature,	55, 70, 85 and 100 °C 1 h	[70]
NanoHAp; Mg-Zn scaffolds	$Ca(NO_3)_2$, $NH_4H_2PO_4$, $NaNO_3$. pH = 5.0	40 mA/cm^2 duty cycle 0.1	85 °C 1 h	[71]
HAp; Au	$Ca(NO_3)_2$. $NH_4H_2PO_4$, H_2O_2, pH 4.5 or 6	Constant mode: 1.6 V vs. Ag/AgCl followed by a pulsed mode duty cycle 0.33	45 min 70°C	[152]
HAp-$Ca_3(PO_4)_2$; Ti6Al4V	$Ca(NO_3)_2$, $NH_4H_2PO_4$, with or without H_2O_2. pH = 4.4	Pulsed mode 8 mA/cm^2	21 min 50 °C	[37]
HAp; nanoTiO$_2$	$Ca(NO_3)_2$, $NH_4H_2PO_4$	2.5 mA/cm^2 Duty cycle 0.5	20–120 s	[49]
CoCa-P; Ti22Nb6Zr	$Ca(NO_3)_2$, $NH_4H_2PO_4$, $Co(NO_3)_2$, H_2O_2.	Pulsed mode 15mA/cm^2	15 min	[87]
FHAp; Mg-Zn-Ca	$NaNO_3$, $NH_4H_2PO_4$ $Ca(NO_3)_2$, NaF, H_2O_2. pH 5.0	Pulse mode 1 mA/cm^2	65 °C	[89]
SiHAp; Mg alloy	$Ca(NO_3)_2$, $NH_4H_2PO_4$, $NaNO_3$, tetraethoxysilane.	Pulse mode 0.4–0.6 V Duty cycle 0.3	40–80°C 40 min	[90]
SrCa-P; Ti6Al4V	$Ca(NO_3)_2$, $NH_4H_2PO_4$, $Sr(NO_3)_2$.	Pulsed mode 15 mA/cm^2	15 min 60 °C	[80]
(Sr,Mg,Zn)HAp; Ti-6Al-4V	$CaCl_2$, $SrCl_2$, $MgCl_2$, $ZnCl_2$, $NH_4H_2PO_4$, H_2O_2. pH 4.5	Pulsed mode 1 mA/cm^2 duty cycle 0.2 and 0.8	1 h 65 °C	[79]
(Sr,Mg)$Ca_3(PO_4)_2$; C/C composite	$Ca(NO_3)_2$ $Sr(NO_3)_2$, $Mg(NO_3)_2$ $NH_4H_2PO_4$.	Pulse mode 2.5 V Duty cycle 0.4.	3 h 50 °C	[92]
(Zn,Mg,Sr,Ag)Ca-P; Ti6Al4V	$Ca(NO_3)_2$, $NH_4H_2PO_4$, $AgNO_3$, $Zn(NO_3)_2$, $Sr(NO_3)_2$, $Mg(NO_3)_2$, H_2O_2.	Pulsed mode 400 mA/cm^2 duty cycle 0.2	70 °C	[88]
ZnHAp	$Ca(NO_3)_2$, $NH_4H_2PO_4$, H_2O_2. pH 4.5	Pulsed mode 0.5–3 mA/cm^2	1 h 65 °C	[95]
HAp + CNTs	$Ca(NO_3)_2$, NH_4H_2PO4, H_2O_2 MWCNTs	3 V	pH 4.7	[95]
Composite Coatings				
Reduced graphene oxide (rGO)-polydopamine-CuNPs-Nil blue; glass carbon	$Cu(NO_3)_2$, phosphate-buffered saline (PBS).	Pulse mode from −0.5 V to 0.8 V, a scan rate of 100 mV/s	-	[95]
MCWNT–HAp; stainless steel	$CaCl_2$, $NH_4H_2PO_4$, NaCl.	Pulsed mode 5, 10, and 20 mA/cm^2 duty cycle 0.2	1 h RT	[103]
HAp-$CaHPO_4$; stainless steel	$CaCl_2$, $NH_4H_2PO_4$.	Pulsed mode Either constant current 5 and 10 mA/cm^2, or constant voltage 1, 2 and 3 V	RT	[58]
Reduced graphene oxide (rGO) and MWCNT/ HAp–calcium orthophosphate phases; stainless steel	$CaCl_2$, $NH_4H_2PO_4$.	Pulsed mode 10 mA/cm^2	900 s RT	[115]
HAp-polypyrrole; stainless steel	$Ca(NO_3)_2$, $NH_4H_2PO_4$, KNO_3, pyrrole monomer.	Pulsed mode 5, 10, and 20 mA/cm^2	1500 s RT	[124]

Table 2. Cont.

Coating; Substrate	Solution	Voltage/Current for Pulsed Mode Deposition	Deposition Time and Temperature	Reference
Composite Coatings				
Graphene oxide (GO)-HAp; Ti	$Ca(NO_3)_2$, $NH_4H_2PO_4$, H_2O_2, GO. pH 4.2	Constant or pulsed mode 15 mA/cm^2 duty cycle 0.1	65 °C	[109]
Graphene oxide (GO)-HAp; Ti	$Ca(NO_3)_2$, $NH_4H_2PO_4$, H_2O_2, GO. pH 4.5	Pulsed mode 15 mA/cm^2 duty cycle 0.1	50 s 65 °C	[110]
GO-calcium phosphate; Ti	$Ca(NO_3)_2$, $NH_4H_2PO_4$, $NaNO_3$, H_2O_2, GO. pH 6	Pulsed mode 15 mA/cm^2 duty cycle 0.1	65 °C	[113]
HAp-CNTs; Mg alloy	$Ca(NO_3)_2$, $NH_4H_2PO_4$, H_2O_2. pH 4.7	−3 V duty cycle 0.2	RT	[114]
Iridium oxide/human plasma proteins	Iridium chloride, oxalic acid, human plasma pH = 10.2	Cyclic voltammetry from −0.6 to 0.8 V vs. Ag/AgCl scan rate 10 mV/s	-	[94]
Polypyrrole/Nb$_2$O$_5$; stainless steel	Pyrrole in oxalic acid, with the addition of Nb$_2$O$_5$	Cyclic voltammetry −0.6 V to +0.7 V vs. SCE the scan rate of 50 mV/s	-	[141]
Polyacrylic acid (PAA) followed by Ga-modified chitosan; Ti	PAA water solution	For PAA only: from 0 to −1.2 V	4 min	[146]
Poly (3,4-ethylenedioxythiophene) (PEDOT)/FHAp; Ti-Nb-Zr	LiClO$_4$, ACN (acetonitrile), monomer EDOT, FHAp	Sweeping the potential from −600 to 1600 mV sweeping rate of 0.05 V/s	-	[123]
ZnO/ZnHAp hybrid coating; carbon fiber	$Zn(NO_3)_2$, $Ca(NO_3)_2$, $NH_4H_2PO_4$.	1st stage: 0.6 mA/cm^2 (ECD). 2nd stage: 3 V EPD.	1st stage: 30 min, 343 K 2nd stage: 60 min	[119]
Zn,Cu/AgNPs	$CuCl_2$, $ZnCl_2$, glycine, cetyltrimethylammonium bromide (CTAB), AgNPs. pH = 10	Cyclic voltammetry from 0.1 to −1.6 V vs.SCE	-	[140]

RT—room temperature. Empty spaces mean that no data have been reported in references.

3.2.1. Effect of an Electrolyte Composition

As for the ECD, applied electrolytes were prepared based on 0.042 mol/L of $Ca(NO_3)_2$ and 0.025 mol/L of $NH_4(H_2PO_4)$. Exceptionally, the SBF solution [58] was used during deposition (NaCl—7.996 g/L, KCl—0.224 g/L, $CaCl_2 \cdot 2H_2O$—0.278 g/L, $MgCl_2 \cdot 6H_2O$—0.305 g/L, $NaHCO_3$—0.350 g/L, $K_2HPO_4 \cdot 3H_2O$—0.228 g/L, and Na_2SO_4—0.071 g/L). The increasing Ca and P contents in an electrolyte affected the morphology of HAp [49], from separated particle to plate-like.

Hydroxyapatite, carbonate apatite, and dicalcium phosphate dehydrate were deposited on carbon/carbon (C/C) composites [76]. The supersaturation degrees of Ca^{2+} and PO_4^{3-} changed the crystalline habit of HAp, led to large crystal size, and transformed crystal shape from belt-like to a plant-like structure. The H_2O_2 was added to prevent hydrogen bubbles formation during electrodeposition and to favor better calcium phosphate crystals nucleation [152]. For HAp coatings deposited on NiTi alloy by PED, the morphology of the coating changed from needle-like to plate-like structure as the electrolyte concentration decreased about five times [64].

3.2.2. Effect of Deposition Potential

The deposition under constant voltage was usually run by cyclic voltammetry between to fixed potential value, come and back. In some reports [152], the pulsed mode was applied, in which the voltage was imposed for 60 s followed by break-time for 120 s. During the deposition of hydroxyapatite, carbonate apatite, and dicalcium phosphate dehydrate on C/C composites [76], the increasing voltage resulted in promoting bioactivity by altering the morphologies and phases. For silver nanoparticles in solution during the electrodeposition of Zn-Cu/AgNP composite coatings by cyclic voltammetry [140], the process was shown to occur through two stages with different energies. The first stage occurred in the potential range from −0.4 to −0.7 V vs. SCE and was mainly associated with the electrodeposition of a copper film, while the second stage corresponded to the bulk deposition of Zn-Cu/AgNPs and occurred from −1.4 to −1.6 V vs. SCE. The increasing HAp nucleation rate vis-à-vis increasing deposition potential was observed, and the presence of calcium orthophosphate phase was decreased along [58] with an increased rate of hydroxyapatite phase in the coating.

3.2.3. Effect of Deposition Current

Using the current pulsed methods, smooth uniform coatings were obtained onto Mg alloy when a current density in a specific range was used (10–30 mA/cm^2) [72]. The growth of particles was observed when low frequencies (50 Hz) and high frequencies (1000 Hz) were applied.

In [58], the effect of current parameters was observed. In other research [91], when Si-HAp coating was deposited on Mg-5Zn-0.3Ca alloy substrate by pulse electrodeposition, at a low current density of 20 mA/cm^2 a coarse and non-uniform coating was deposited, and at a high current density of 60 mA/cm^2, high amounts of hydrogen gas at the interface were produced, and a non-uniform coating was formed. At the middle current density of 40 mA/cm^2, nano-needle like coating was observed. For higher peak current density, the Ca-P coating consisting of needle-shaped crystals was formed with pores in between [54]. At lower current density of 0.5 mA/cm^2 a more compact uniform plate-like morphology was observed than those at a higher current density of 3 mA/cm^2.

In [152], by the template-assisted pulsed electrodeposition method, the calcium-deficient hydroxyapatite (CDHA) particles in aqueous baths with hydrogen peroxide were used by both applying pulsed current density and pulsed potential in cathodic electrodeposition. Deposition in the membrane was successful only for higher values of current density, so a valid value of 120–250 mA/cm^2 was applied. In other research [101], a composite coating of hydroxyapatite and calcium hydrogen phosphate over 316L grade of stainless steel performed by PED at 10 mA/cm^2 demonstrated the highest weight percentage and crystallinity of hydroxyapatite phase and a continuous, faster and interconnected cell growth. For a composite coating of hydroxyapatite-polypyrrole, the coating deposited with moderate current density (10 mA/cm^2) seems to be the optimum one regarding the faster-interconnected growth of MG63 cells over the coating surface along with highest corrosion resistance and anodic passivation capability [124]. In [65], the electrodeposition at the higher current densities of 15 and 20 mA/cm^2 increased the possibility of the hydroxyapatite phase formation in the coating rather than the other less stable calcium phosphate phases. The optimum conditions to create a uniform nano-hydroxyapatite coating on the Mg-Zn scaffold was found at 40 mA/cm^2 and 0.1 duty cycle [71]. Finally, in research [63], in which reverse current densities of 0.1 mA/cm^2 were applied, the porosity was increased by increasing the current density of the reverse pulse. Also, in this condition, the resulted film was composed of nanosized HA crystals. It seems that the reverse current step can influence the porosity and the structure of the coatings by the dissolution of the unstable phases, which are formed during the direct current stage of the bipolar pulse deposition. For the production of acceptable quality Ca-P coatings on Mg alloy, the current density should be between 10 and 30 mA/cm^2 [72]. In [49], the observed HAp morphologies consisted of particles, a mixture of particles and plate-like shapes, and entirely plate-like shapes as related to the increasing number of deposition cycles.

3.2.4. Effect of Deposition pH

The pH value was generally acidic. For deposited SiHAp coating at pH 5, the needle-like morphology, but the coatings deposited at pH 4 and 6, the plate-like morphology with micro-size thickness were observed [91].

3.2.5. Effect of Deposition Temperature

The applied were either room temperature or elevated temperatures, like 40–60 °C [53], 85 °C [70], or even 200 °C [25] temperatures were reported. In [91], tests were made at temperatures ranging from 25 to 100 °C, at temperatures up to 85 °C, the morphology of the SiHAp coating changed to the nano-sized needle-like blades.

3.3. Electrophoretic Deposition (EPD)

Table 3 illustrates the process determinants of several investigations performed by EPD.

Table 3. The main determinants of the electrophoretic deposition of biocoatings.

Coating; Substrate	Electrolyte	Voltage/Current	Deposition Time and Temperature	Reference
HAp; NiTi	ethanol	40 V	20 s	[39]
HAp; TiO$_2$	-	200 V	1 min	[44]
HAp; Ti	ethanol	10–45 V	1–8 min	[42]
HAp; TiO$_2$	ethanol	60 V	45 s RT	[45]
HAp; Au	ethanol and octadecyltri-chlorosilane	70 V/cm^2	1 h	[131]
Hap + TiO$_2$	acetylacetone	20 V	30–120 s	[43]
HAp + CNTs; NiTi	ethanol	30 V	30 s	[116]
Hap + MWCNTs	butanol	60 V	2 min	[117]
Hap + MNWCNTs + nanoAg + nanoCu	ethanol, isopropanol	11 and 30 V	2 min RT	[138]
nanoHAp; Ag	ethanol	10 V		[47]
Nano(Zn/Ca)HAp; (Si)Ti	-	10, 50, and 100 V/cm	1 min pH 12	[93]
nanoHAp; Ti	ethanol	10 V	10 min	[41]
nanoHAp; (Mg,Zr,Ce) oxides	water	380 V	10 min	[46]
nanoHAp	0.1, 0.2 or 0.5 g nanoHAp	15, 30, and 50 V	1 min	[40]
nanoHAp + nanoAg	ethanol	15 and 30 V for nanoHAp 60 V for nanoAg	1 min for nanoHAp 5 min for nanoAg	[135]
nanoHAp + nanoAg	ethanol	50 V	1 min	[136]
nanoHAp + nanoCu; TiO2	ethanol	30 V	1 and 2 min RT	[137]
nanoHAp + borium nitride; Ti	ethanol	100 and 150 V	5–20 s pH 4	[120]
PEEK + HAp	ethanol	75 V	45 s pH 5.5	[130]
HAp+Si + MWCNTs	-	cathodic	-	[153]
Chitosan + HAp; TiO$_2$	acetic acid, etanol, water	10–15 V	3–9 min	[154]
Hap + ZNHAp + MWCNTs + chitosan	-	-	-	[128]
Bioglass + HAp (whiskers)	isopropanol	40 V	1 min	[121]
Chitosan + bioglass + HAp	acetic acid, etanol, water	20 and 30 V	5 and 15 min pH 3.3, 4, 5	[155]
GO (graphene oxide) + MgHAp; C/C composites	-	-	-	[111]
The poly-l-lysine (PLL) + 3,4-dihydroxybenzylaldehyde (DHBA) + HAp + TiO$_2$	ethanol-water	50 V	-	[129]
methacrylates	dioxan	15 V	5 min	[99]
PEEK	ethanol	70–115 V	1 min	[100]
Chitosan + Eudragit	acetic acid	10 and 30 V	1 and 3 min	[145]
PMMA + soy lecithin	the microemulsion of coconut oil and water	1 mA (4–15 V)	30 min RT	[147]

RT—room temperature. Empty spaces mean that no data have been reported in references.

3.3.1. Effect of an Electrolyte Composition

The electrolyte composition affected biological properties. Galindo et al. [93] studied the effect of different Zn/Ca ratio on cytocompatibility and antibacterial efficiency demonstrating the best results at the 5% and 10% content of Zn in the carbonated hydroxyapatite (CHAp). In-vitro bioactivity of PEEK-HAp coatings was enhanced by increasing the amount of HAp particles, but such coatings had inferior adhesion to the substrate [130].

The electrolyte also affected corrosion resistance, adhesion, and wettability. The corrosion current, between 1 and 10 nA/cm^2, was significantly higher for undecorated nanoHAp coatings and close to that of the substrate for decorated nanoHAp coatings [135]. The compact structure of hydroxyapatite with 20 wt.% silicon and hydroxyapatite with 20 wt.% silicon-1 wt.% multi-walled carbon nano-tubes coatings could efficiently increase the corrosion resistance of NiTi substrate [153].

The presence of nanosilver particles in the coating improved adhesive properties and reduce contact angle values [136]. Besides, the presence of nanosilver particles has a significant effect on homogeneity and quality of coating. In the presence of the silver nanoparticles, a smaller number of cracks and a smaller dimension of cracks were observed [40]. On the other hand, in [120] the increase in the amount of BN was insignificant as concerns the coating thickness.

In [121], the electrophoretic deposition of bioactive glass (BG) coatings reinforced by whisker hydroxyapatite (WHA) particles at 50 and 75 wt.% WHA showed the highest bonding strength and bioactivity response. Addition of other ceramics, MWCNTs, changed mechanical properties [117]: the hardness and adhesion strength of the HA coating improved from 72 HV and 17.2 MPa to 405 HV and 32.1 MPa. Moreover, the results of the biological tests revealed an incredible improvement in the apatite and bone cell growth on the HA coatings composed of titanium and MWCNTs. For a duplex coating composed of electrophoretically deposited graphene oxide (GO), inner-layer and electrodeposited GO/Mg substituted hydroxyapatite (MH) outer-layer on carbon/carbon composites (CC) demonstrated the bonding strength between the duplex coating and CC as of 7.4 MPa, about 80% higher than without GO [111]. The composition of HAp and MWCNTs promoted also the osteoblast adhesion and proliferation [129] and bioactivity [116].

The size of HPa seems to play an important role in biological behavior; the cell adhesion was better on the small- and middle-sized nanocrystals (SHA, MHA > LHA), which was interestingly opposite in cell spreading (LHA > SHA, MHA), and then the cell proliferation was again up-regulated on SHA and MHA [42].

3.3.2. Effect of Deposition Potential

In [42], the increasing potential at EPD of nanoHAp resulted in increasing coating thickness, the release rate of silver in SBF, resistance to scratch, but in decreasing hardness. On the contrary, in other reports [120], the voltage effect was insignificant as concerns the thickness.

3.3.3. Effect of Deposition Time

The increase in deposition time results in an increased thickness, decreasing hardness and increasing adhesion strength, and significantly decreasing the contact angle in wettability tests [137]. When EPD was made on the nanotubular TiO_2 surface, at a shorter 1 min EPD time, hydroxyapatite was deposited inside the nanotubes, between nanotubes, and also formed a thin layer over the nanotubes. At a longer 2 min, EPD time, the layer of HAp over the HAp-TiO_2 composite was thicker. The increasing thickness with deposition time was observed earlier [120].

3.4. Plasma Electrochemical Oxidation (PEO)

The plasma electrochemical oxidation, also called micro-arc oxidation (MAO), is generally assumed as a tool to generate a surface with high roughness affecting the adhesion of osteoblasts. However, PEO, when made in an electrolyte composed of different compounds, may significantly change the surface chemical and phase composition. Therefore, the attempts were made to obtain bioactive and well-adjacent to the substrate coatings using this approach.

The significant part of such research was performed on titanium substrates. In [156], the coatings, consisting of HAp and titanium oxide, were produced on Ti6Al4V alloy in a solution containing calcium acetate and β-calcium glycerophosphate. The phases of anatase, rutile, TCP ($Ca_3(PO_4)_2$), perovskite-$CaTiO_3$, and HAp were detected, their crystallinity more profound with increasing treatment time. The friction and wear resistance properties, and corrosion resistance of the PEO coatings were substantially improved compared to Ti6Al4V alloy. In similar research [157], the HAp/TiO_2 composite coating was prepared on a titanium surface. The flocculent structures were obtained during the early stages of treatment and, as the treatment period extended, increasing amounts of Ca-P precipitate appeared on the surface, and the flocculent morphology transformed into a plate-like morphology, and finally into flower-like apatite. The Ca/P atomic ratio gradually decreased, and the adhesive strength between the apatite and TiO_2 coating was improved. The search for a new complexing agent was described in [158]. The barium titanate coatings were also obtained [159] on Ti6Al4V alloy, and the coatings presented lower friction coefficient, higher wear, and corrosion resistance than uncoated Ti alloys. In [160], the coatings formed in a solution containing Ca and P achieved a thickness of 11.78 μm, and bond strength 33.69 MPa. Recently [161], two-layer hydroxyapatite CuHAp-TiO_2 composite coating was prepared on titanium in an aqueous electrolyte containing Ca, P, and Cu at a constant current density of 100 mA/cm^2. Yu et al. [48] obtained Mg-containing hydroxyapatite coatings on Ti-6Al-4V alloy for dental materials by plasma electrolytic oxidation. The increasing Mg content in an electrolyte resulted in decreasing the irregularity of the surface, pore size, and the number of pores decreased as the Mg concentration increased.

After 20 min of process, Cu species incorporated into the TiO_2 as CuO and Cu_2O, resulting in the formation of Cu-doped TiO_2-based coating. Dziaduszewska et al. [162] found the best process parameters in a similar electrolyte as of 300 V for 15 min, at 32 or 50 mA of current value, which resulted in the proper microstructure, high Ca/P ratio, hydrophilicity, early-stage bioactivity. The coating mechanical properties, Young's modulus, and hardness were close to the values characteristic for bones, and adhesion was also better than at 200 V of voltage.

In addition to HAp coatings on Ti, also other substrates were covered with HAp for biological applications. The HAp coatings were grown on Ta samples by PEO at the potential in a range of 350 to 500 V, and treatment time from 60 to 600 s [163]. The formation of crystalline HAp was detected for samples treated over 180 s at 500 V. In [164], the coating was obtained by PEO on Zr in for process duration of 2 to 30 min. The increasing oxidation resulted in thicker and rougher coatings. The increasing oxidation resulted in thicker and rougher coatings. In [165], a hydrophilic surface was obtained by PEO on Mg with a biosynthesized hyaluronic acid and carboxymethylcellulose to improve bonding function.

3.5. Electro-Spark Deposition (ESD)

The ESD method was carried out on Ti6Al4V to deposit a layer on the steel substrates at the first step, and then the PEO process was employed to improve properties of the alloyed titanium layer at the second step [166]. The duplex coating consisted of α-Al_2O_3 (corundum) and γ-Al_2O_3 phases were formed after the PEO process while $AlFe_3$, TiN, and $AlTi_3$ phases were detected in the ESD coating. In the next research [167], the HAp coatings were formed on steel again by ESD and PEO. The HAp-based surface was rough and porous, indicated a hydrophilic character, and improved bioactivity in SBF.

3.6. Electro-Discharge Method (EDM)

In [168], the hydroxyapatite coating was synthesized on the Zr-based bulk metallic glass surface to promote cell proliferation. The formation of the nanoporous coating of about 27.2 μm thick appeared, firmly adhered to adhesive failure at 132 N. A significant rise in cell viability was also shown. In [169], the surface modification of the β-type titanium substrate was made by EDT to incorporate MWCNTs and μ-hydroxyapatite (μHAp) powders in the dielectric medium to promote cell proliferation.

3.7. Electropolymerization (EP)

The electropolymerization is a method applied for some organic coatings. In [170], MWCNTs–polyaniline composite films were prepared by in situ electrochemical polymerizations on the titanium from an aniline solution containing a small content of well-dispersed CNTs. Then these composite films were employed as a substrate for the electrodeposition of platinum nanoparticles forming the surface highly active for electrocatalytic oxidation of glycerol. In a more recent work [171], the polypyrrole (PPy)/chitosan composite coating on titanium was obtained by EP method showing enhanced microhardness and adhesion strength and higher corrosion resistance compared to the PPy coating.

Some specific environments were also utilized. In [172], the enzyme laccase was immobilized during the potentiostatic deposition of a thin polydopamine film (PDA) on carbon surfaces.

3.8. Effects of Electrodeposition Method on Properties of Coatings

The electrodeposition methods are important, even not prevalent (the sprayed coatings are the mostly applied for joint implants) for coatings on orthopedic and dental implants, maxillofacial implants, and stents. However, the choice of method depends on the designed coating, its predicted properties, and applications.

The most popular are electrocathodic deposition at either constant or pulse voltage/current, and as the third, electrophoretic deposition. The same coatings, mainly ceramic or composite ceramics-ceramics, ceramics-polymer, or ceramics-metal can be obtained by each of these methods. Therefore, the research was conducted on similar coatings by these methods even if there were few attempts to compare the coatings of the identical composition obtained by all EDM techniques.

ECD is the simplest and most developed technique in the past. However, it is excellent only for some metals. As concerns ceramics which cannot dissociate to ions, their deposition on the metal base is difficult, the surfaces are often non-stoichiometric and non-homogenous. Despite that, it is still the most popular deposition method, thanks to its simplicity, and it is widely used for Ca-P and HAp coatings, one or multicomponent layers. The applied voltage is between −1.5 to 2.5 V (SCE) for the Ti based materials and a little more for Mg alloys. The process is usually made under voltage control, and for current control, the current values are 0.1–3 mA/cm^2. Deposition time depends on a used compound, and it is very short for metals, 15–60 s, and much longer for ceramics, For Ca-P and HAp coatings, even up to 1 h. The used bath is based mainly on two substances, $Ca(NO_3)_2$ and $NH_4H_2PO_4$. The process parameters seem to play a less important role: the electrolyte composition is determined by the ionic components of designed coating, the voltage must be maintained within some limits, pH is usually neutral or slightly acidic. The ECD gives then no great possibility to adjust the properties of coatings by process parameters.

The pulse electrocathodic deposition (PED) gives more possibilities to control the coating. The voltage or, more seldom, current, is the obvious process parameter, but a duty cycle and number of cycles (deposition time) allow creating of previously design coating. The electric parameters are like those used under direct current, the duty cycle is between 0.1 and 0.5, deposition time reaches even 1 h.

The electrophoretic deposition is exceptionally plausible for no dissociated compounds such as ceramics and polymers. For the PED, process parameters such as duty cycle and the number of cycles need some preliminary tests. For the EPD, the most significant difficulty is to use the proper bath components (together with their contents in a bath) and solvent, in which a stable suspension is obtained and the species possess the proper charge. Ethanol, butanol, and isopropanol are mainly applied, but this list can likely be extended in the future affecting the process efficiency and coating adhesion and thickness. It is noteworthy that for the attempts to look for the most optimum electrochemical conditions: voltage was changed between 10 and 100 V, and once even it was 380 V, and deposition time was usually short, between 15 s and 2 min, sometimes longer. The use of such a short time allows obtaining the thin coatings, particularly suitable for coatings on metallic implants, for which an excessive thickness may result in a weak adhesion and failure. Therefore, it is a great advantage of EPD.

The critical factor for adhesion is the surface roughness. As a rule, it is adjusted by mechanical polishing, chemical etching, oxidation of Ti to nanotubular structure, and others. However, the micro-arc oxidation, MAO (called also plasma electrochemical oxidation, PEO) is the unique among electrodeposition methods, as it brings out the rough surface rich in biological elements due to high voltage oxidation in electrolytes containing Ca and P compounds. For that reason, it is commonly used for dental implants.

Other methods, such as electro-spark, electrodeposition, and electropolymerization techniques are less used. Despite that, they can be more intensively developed in the future as they allow creating the duplex coatings, nanoporous coatings, and polymer coatings.

4. Conclusions

The coatings proposed for medical applications are, to a greater extent, either ceramic or ceramic–ceramic or ceramic–polymer and, to a lesser extent, either polymer, metallic, and other composite coatings such as ceramic–metallic, ceramic–polymer, polymer-polymer, and others.

The most applied compounds are phosphates, and the most used substrates are titanium and its alloys and, to a lesser extent, NiTi alloys.

Three electrocrystallization methods are preferred for deposition of biocoatings such as direct and pulse electrocathodic deposition, and electrophoretic deposition. Plasma electrochemical oxidation, electro-spark, electro-discharge, and electropolymerization are not often applied.

The main process determinants include potential or current, and electrolyte composition, more seldom deposition time, pH, and temperature.

The current progress in biocoatings focuses on an enhancement of their bioactivity, biocompatibility (hydrophilicity), adhesion to substrates, mechanical strength, and corrosion resistance. Such properties are determined by microstructure and thickness of coatings, and characteristics of the interface.

Future research is to be directed towards the development of biocoatings possessing efficient and long term antibacterial efficiency, with no short and long term cytotoxicity, enhancing adhesion, proliferation, and viability of cells, well adjacent to the substrate, while being resistant to shear and compressive stresses, environmentally friendly, and economically justified.

Author Contributions: Conceptualization, A.Z. and M.B.; formal analysis, A.Z.; investigation, A.Z. and M.B.; writing—original draft preparation, A.Z. and M.B. All authors have read and agreed to the published version of the manuscript.

Funding: This research received no external funding.

Conflicts of Interest: The authors declare no conflict of interest.

References

1. Nakaya, M.; Uedono, A.; Hotta, A. Recent progress in gas barrier thin film coatings on PET bottles in food and beverage applications. *Coatings* **2015**, *5*, 987–1001. [CrossRef]
2. Aranke, O.; Algenaid, W.; Awe, S.; Joshi, S. Coatings for automotive gray cast iron brake discs: A review. *Coatings* **2019**, *9*, 552. [CrossRef]
3. Ma, L.; Eom, K.; Geringer, J.; Jun, T.S.; Kim, K. Literature review on fretting wear and contact mechanics of tribological coatings. *Coatings* **2019**, *9*, 501. [CrossRef]
4. Verbič, A.; Gorjanc, M.; Simončič, B. Zinc oxide for functional textile coatings: Recent advances. *Coatings* **2019**, *9*, 550. [CrossRef]
5. Bir, F.; Khireddine, H.; Touati, A.; Sidane, D.; Yala, S.; Oudadesse, H. Electrochemical depositions of fluorohydroxyapatite doped by Cu^{2+}, Zn^{2+}, Ag^+ on stainless steel substrates. *Appl. Surf. Sci.* **2012**, *258*, 7021–7030. [CrossRef]

6. Tiwari, A.; Seman, S.; Singh, G.; Jayaganthan, R. Nanocrystalline cermet coatings for erosion-corrosion protection. *Coatings* **2019**, *9*, 400. [CrossRef]
7. Mandracci, P.; Mussano, F.; Rivolo, P.; Carossa, S. Surface treatments and functional coatings for biocompatibility improvement and bacterial adhesion reduction in dental implantology. *Coatings* **2016**, *6*, 7. [CrossRef]
8. Hou, N.Y.; Perinpanayagam, H.; Mozumder, M.S.; Zhu, J. Novel development of biocompatible coatings for bone implants. *Coatings* **2015**, *5*, 737–757. [CrossRef]
9. Graziani, G.; Boi, M.; Bianchi, M. A review on ionic substitutions in hydroxyapatite thin films: Towards complete biomimetism. *Coatings* **2018**, *8*, 269. [CrossRef]
10. Cometa, S.; Bonifacio, M.A.; Mattioli-Belmonte, M.; Sabbatini, L.; De Giglio, E. Electrochemical strategies for titanium implant polymeric coatings: The why and how. *Coatings* **2019**, *9*, 268. [CrossRef]
11. Sartori, M.; Maglio, M.; Tschon, M.; Aldini, N.N.; Visani, A.; Fini, M. Functionalization of ceramic coatings for enhancing integration in osteoporotic bone: A systematic review. *Coatings* **2019**, *9*, 312. [CrossRef]
12. Duta, L.; Popescu, A.C. Current status on pulsed laser deposition of coatings from animal-origin calcium phosphate sources. *Coatings* **2019**, *9*, 335. [CrossRef]
13. Paital, S.R.; Dahotre, N.B. Calcium phosphate coatings for bio-implant applications: Materials, performance factors, and methodologies. *Mater. Sci. Eng. R Rep.* **2009**, *66*, 1–70. [CrossRef]
14. Guslitzer-Okner, R.; Mandler, D. Electrochemical coating of medical implants. In *Applications of Electrochemistry and Nanotechnology 291 in Biology and Medicine I*; Eliaz, N., Ed.; Springer Science & Business Media: Berlin, Germany, 2011; pp. 291–342. ISBN 9781461403470.
15. Kulkarni, M.; Mazare, A.; Schmuki, P.; Iglič, A. Biomaterial surface modification of titanium and titanium alloys for medical applications. In *Nanomedicine*; Seifalian, A., de Mel, A., Kalaskar, D.M., Eds.; One Central Press Altrincham: Cheshire, UK, 2014; pp. 111–136.
16. Asri, R.I.M.; Harun, W.S.W.; Hassan, M.A.; Ghani, S.A.C.; Buyong, Z. A review of hydroxyapatite-based coating techniques: Sol-gel and electrochemical depositions on biocompatible metals. *J. Mech. Behav. Biomed. Mater.* **2016**, *57*, 95–108. [CrossRef]
17. Dorozhkin, S.V. Calcium orthophosphates (CaPO$_4$): Occurrence and properties. *Prog. Biomater.* **2016**, *5*, 9–70. [CrossRef]
18. Adeleke, S.A.; Bushroa, A.R.; Sopyan, I. Recent development of calcium phosphate-based coatings on titanium alloy implants. *Surf. Eng. Appl. Electrochem.* **2017**, *53*, 419–433. [CrossRef]
19. Liu, W.; Liu, S.; Wang, L. Surface modification of biomedical titanium alloy: Micromorphology, microstructure evolution and biomedical applications. *Coatings* **2019**, *9*, 249. [CrossRef]
20. Su, Y.; Cockerill, I.; Zheng, Y.; Tang, L.; Qin, Y.X.; Zhu, D. Biofunctionalization of metallic implants by calcium phosphate coatings. *Bioact. Mater.* **2019**, *4*, 196–206. [CrossRef]
21. Yang, J.; Cui, F.; Lee, I.S. Surface modifications of magnesium alloys for biomedical applications. *Ann. Biomed. Eng.* **2011**, *39*, 1857–1871. [CrossRef]
22. Wan, P.; Tan, L.; Yang, K. Surface modification on biodegradable magnesium alloys as orthopedic implant materials to improve the bio-adaptability: A review. *J. Mater. Sci. Technol.* **2016**, *32*, 827–834. [CrossRef]
23. Ananthi, A.; Kumar, S.S.; Phani, K.L. Facile one-step direct electrodeposition of bismuth nanowires on glassy carbon electrode for selective determination of folic acid. *Electrochim. Acta* **2015**, *151*, 584–590. [CrossRef]
24. Cassar, I.R.; Yu, C.; Sambangi, J.; Lee, C.D.; Whalen, J.J.; Petrossians, A.; Grill, W.M. Electrodeposited platinum-iridium coating improves in vivo recording performance of chronically implanted microelectrode arrays. *Biomaterials* **2019**, *205*, 120–132. [CrossRef] [PubMed]
25. Arnould, C.; Delhalle, J.; Mekhalif, Z. Multifunctional hybrid coating on titanium towards hydroxyapatite growth: Electrodeposition of tantalum and its molecular functionalization with organophosphonic acids films. *Electrochim. Acta* **2008**, *53*, 5632–5638. [CrossRef]
26. Frank, M.J.; Walter, M.S.; Tiainen, H.; Rubert, M.; Monjo, M.; Lyngstadaas, S.P.; Haugen, H.J. Coating of metal implant materials with strontium. *J. Mater. Sci. Mater. Med.* **2013**, *24*, 2537–2548. [CrossRef]
27. Mollamahale, Y.B.; Ghorbani, M.; Dolati, A.; Hosseini, D. Electrodeposition of well-defined gold nanowires with uniform ends for developing 3D nanoelectrode ensembles with enhanced sensitivity. *Mater. Chem. Phys.* **2018**, *213*, 67–75. [CrossRef]

28. Nasirpouri, F.; Cheshideh, H.; Samardak, A.Y.; Ognev, A.V.; Zubkov, A.A.; Samardak, A.S. Morphology- and magnetism-controlled electrodeposition of Ni nanostructures on TiO$_2$ nanotubes for hybrid Ni/TiO$_2$ functional applications. *Ceram. Int.* **2019**, *45*, 11258–11269. [CrossRef]
29. Pavez, J.; Silva, J.F.; Melo, F. Homogeneous calcium carbonate coating obtained by electrodeposition: In situ atomic force microscope observations. *Electrochim. Acta* **2005**, *50*, 3488–3494. [CrossRef]
30. Park, J.H.; Lee, D.Y.; Oh, K.T.; Lee, Y.K.; Kim, K.M.; Kim, K.N. Bioactivity of calcium phosphate coatings prepared by electrodeposition in a modified simulated body fluid. *Mater. Lett.* **2006**, *60*, 2573–2577. [CrossRef]
31. Park, J.H.; Lee, Y.K.; Kim, K.M.; Kim, K.N. Bioactive calcium phosphate coating prepared on H$_2$O$_2$-treated titanium substrate by electrodeposition. *Surf. Coat. Technol.* **2005**, *195*, 252–257. [CrossRef]
32. Yang, X.; Zhang, B.; Lu, J.; Chen, J.; Zhang, X.; Gu, Z. Biomimetic Ca-P coating on pre-calcified Ti plates by electrodeposition method. *Appl. Surf. Sci.* **2010**, *256*, 2700–2704. [CrossRef]
33. Lin, D.Y.; Wang, X.X. Preparation of hydroxyapatite coating on smooth implant surface by electrodeposition. *Ceram. Int.* **2011**, *37*, 403–406. [CrossRef]
34. Mokabber, T.; Zhou, Q.; Vakis, A.I.; van Rijn, P.; Pei, Y.T. Mechanical and biological properties of electrodeposited calcium phosphate coatings. *Mater. Sci. Eng. C* **2019**, *100*, 475–484. [CrossRef] [PubMed]
35. Harun, W.S.W.; Asri, R.I.M.; Alias, J.; Zulkifli, F.H.; Kadirgama, K.; Ghani, S.A.C.; Shariffuddin, J.H.M. A comprehensive review of hydroxyapatite-based coatings adhesion on metallic biomaterials. *Ceram. Int.* **2018**, *44*, 1250–1268. [CrossRef]
36. Gopi, D.; Karthika, A.; Sekar, M.; Kavitha, L.; Pramod, R.; Dwivedi, J. Development of lotus-like hydroxyapatite coating on HELCDEB treated titanium by pulsed electrodeposition. *Mater. Lett.* **2013**, *105*, 216–219. [CrossRef]
37. Drevet, R.; Fauré, J.; Sayen, S.; Marle-Spiess, M.; El Btaouri, H.; Benhayoune, H. Electrodeposition of biphasic calcium phosphate coatings with improved dissolution properties. *Mater. Chem. Phys.* **2019**, *236*, 1–7. [CrossRef]
38. Fathyunes, L.; Khalil-Allafi, J. Effect of employing ultrasonic waves during pulse electrochemical deposition on the characteristics and biocompatibility of calcium phosphate coatings. *Ultrason. Sonochem.* **2018**, *42*, 293–302. [CrossRef]
39. Dudek, K.; Dulski, M.; Goryczka, T.; Gerle, A. Structural changes of hydroxyapatite coating electrophoretically deposited on NiTi shape memory alloy. *Ceram. Int.* **2018**, *44*, 11292–11300. [CrossRef]
40. Bartmanski, M.; Zielinski, A.; Majkowska-Marzec, B.; Strugala, G. Effects of solution composition and electrophoretic deposition voltage on various properties of nanohydroxyapatite coatings on the Ti13Zr13Nb alloy. *Ceram. Int.* **2018**, *44*, 19236–19246. [CrossRef]
41. Jażdżewska, M.; Majkowska-Marzec, B. Hydroxyapatite deposition on the laser modified Ti13Nb13Zr alloy. *Adv. Mater. Sci.* **2018**, *17*, 5–13. [CrossRef]
42. Patel, K.D.; Singh, R.K.; Lee, J.H.; Kim, H.W. Electrophoretic coatings of hydroxyapatite with various nanocrystal shapes. *Mater. Lett.* **2019**, *234*, 148–154. [CrossRef]
43. Araghi, A.; Hadianfard, M.J. Fabrication and characterization of functionally graded hydroxyapatite/TiO$_2$ multilayer coating on Ti-6Al-4V titanium alloy for biomedical applications. *Ceram. Int.* **2015**, *41*, 12668–12679. [CrossRef]
44. Albayrak, O.; El-Atwani, O.; Altintas, S. Hydroxyapatite coating on titanium substrate by electrophoretic deposition method: Effects of titanium dioxide inner layer on adhesion strength and hydroxyapatite decomposition. *Surf. Coat. Technol.* **2008**, *202*, 2482–2487. [CrossRef]
45. Pavlović, M.R.P.; Eraković, S.G.; Pavlović, M.M.; Stevanović, J.S.; Panić, V.V.; Ignjatović, N.L. Anaphoretical/oxidative approach to the in-situ synthesis of adherent hydroxyapatite/titanium oxide composite coatings on titanium. *Surf. Coat. Technol.* **2019**, *358*, 688–694. [CrossRef]
46. Xiong, Y.; Lu, C.; Wang, C.; Song, R. Degradation behavior of n-MAO/EPD bio-ceramic composite coatings on magnesium alloy in simulated body fluid. *J. Alloy. Compd.* **2015**, *625*, 258–265. [CrossRef]
47. Fernando, N.L.; Kottegoda, N.; Jayanetti, S.; Karunaratne, V.; Jayasundara, D.R. Stability of nano-hydroxyapatite thin coatings at liquid/solid interface. *Surf. Coat. Technol.* **2018**, *349*, 24–31. [CrossRef]
48. Yu, J.M.; Choe, H.C. Mg-containing hydroxyapatite coatings on Ti-6Al-4V alloy for dental materials. *Appl. Surf. Sci.* **2018**, *432*, 294–299. [CrossRef]
49. Jeong, Y.H.; Kim, E.J.; Brantley, W.A.; Choe, H.C. Morphology of hydroxyapatite nanoparticles in coatings on nanotube-formed Ti-Nb-Zr alloys for dental implants. *Vacuum* **2014**, *107*, 297–303. [CrossRef]

50. Bucur, A.I.; Linul, E.; Taranu, B.O. Hydroxyapatite coatings on Ti substrates by simultaneous precipitation and electrodeposition. *Appl. Surf. Sci.* **2020**, *527*, 146820. [CrossRef]
51. Sun, X.; Lin, H.; Zhang, C.; Jin, J.; Di, S. Electrochemical studies on CaP electrodeposition on three dimensional surfaces of selective laser melted titanium scaffold. *Coatings* **2019**, *9*, 667. [CrossRef]
52. Parcharoen, Y.; Kajitvichyanukul, P.; Sirivisoot, S.; Termsuksawad, P. Hydroxyapatite electrodeposition on anodized titanium nanotubes for orthopedic applications. *Appl. Surf. Sci.* **2014**, *311*, 54–61. [CrossRef]
53. Vidal, E.; Buxadera-Palomero, J.; Pierre, C.; Manero, J.M.; Ginebra, M.P.; Cazalbou, S.; Combes, C.; Rupérez, E.; Rodríguez, D. Single-step pulsed electrodeposition of calcium phosphate coatings on titanium for drug delivery. *Surf. Coat. Technol.* **2019**, *358*, 266–275. [CrossRef]
54. Gopi, D.; Indira, J.; Kavitha, L. A comparative study on the direct and pulsed current electrodeposition of hydroxyapatite coatings on surgical grade stainless steel. *Surf. Coat. Technol.* **2012**, *206*, 2859–2869. [CrossRef]
55. Thanh, D.T.M.; Nam, P.T.; Phuong, N.T.; Que, L.X.; Van Anh, N.; Hoang, T.; Lam, T.D. Controlling the electrodeposition, morphology and structure of hydroxyapatite coating on 316L stainless steel. *Mater. Sci. Eng. C* **2013**, *33*, 2037–2045. [CrossRef] [PubMed]
56. Büyüksağiş, A.; Bulut, E.; Kayalı, Y. Corrosion behaviors of hydroxyapatite coated by electrodeposition method of Ti6Al4V, Ti and AISI 316L SS substrates. *Prot. Met. Phys. Chem. Surf.* **2013**, *49*, 776–787. [CrossRef]
57. Chakraborty, R.; Sengupta, S.; Saha, P.; Das, K.; Das, S. Synthesis of calcium hydrogen phosphate and hydroxyapatite coating on SS316 substrate through pulsed electrodeposition. *Mater. Sci. Eng. C* **2016**, *69*, 875–883. [CrossRef]
58. Chakraborty, R.; Saha, P. A comparative study on surface morphology and electrochemical behaviour of hydroacxyapatite-calcium hydrogen phosphate composite coating synthesized in-situ through electro chemical process under various deposition conditions. *Surf. Interfaces* **2018**, *12*, 160–167. [CrossRef]
59. Shibli, S.M.A.; Jayalekshmi, A.C. A novel nano hydroxyapatite-incorporated Ni-P coating as an effective inter layer for biological applications. *J. Mater. Sci. Mater. Med.* **2009**, *20*, 711–718. [CrossRef]
60. Lin, D.Y.; Wang, X.X. Electrodeposition of hydroxyapatite coating on CoNiCrMo substrate in dilute solution. *Surf. Coat. Technol.* **2010**, *204*, 3205–3213. [CrossRef]
61. Coşkun, M.I.; Karahan, I.H.; Yücel, Y. Optimized electrodeposition concentrations for hydroxyapatite coatings on CoCrMo biomedical alloys by computational techniques. *Electrochim. Acta* **2014**, *150*, 46–54. [CrossRef]
62. Etminanfar, M.R.; Khalil-Allafi, J.; Montaseri, A.; Vatankhah-Barenji, R. Endothelialization and the bioactivity of Ca-P coatings of different Ca/P stoichiometry electrodeposited on the Nitinol superelastic alloy. *Mater. Sci. Eng. C* **2016**, *62*, 28–35. [CrossRef]
63. Etminanfar, M.R.; Khalil-Allafi, J.; Parsa, A.B. On the electrocrystallization of pure hydroxyapatite nanowalls on Nitinol alloy using a bipolar pulsed current. *J. Alloy. Compd.* **2016**, *678*, 549–555. [CrossRef]
64. Marashi-Najafi, F.; Khalil-Allafi, J.; Etminanfar, M.R. Biocompatibility of hydroxyapatite coatings deposited by pulse electrodeposition technique on the Nitinol superelastic alloy. *Mater. Sci. Eng. C* **2017**, *76*, 278–286. [CrossRef] [PubMed]
65. Sheykholeslami, S.O.R.; Khalil-Allafi, J.; Fathyunes, L. Preparation, characterization, and corrosion behavior of calcium phosphate coating electrodeposited on the modified nanoporous surface of NiTi alloy for biomedical applications. *Metall. Mater. Trans. A Phys. Metall. Mater. Sci.* **2018**, *49*, 5878–5887. [CrossRef]
66. Song, Y.W.; Shan, D.Y.; Han, E.H. Electrodeposition of hydroxyapatite coating on AZ91D magnesium alloy for biomaterial application. *Mater. Lett.* **2008**, *62*, 3276–3279. [CrossRef]
67. Zhang, C.Y.; Zeng, R.C.; Chen, R.S.; Liu, C.L.; Gao, J.C. Preparation of calcium phosphate coatings on Mg-1.0Ca alloy. *Trans. Nonferrous Met. Soc. China Engl. Ed.* **2010**, *20*, 655–659. [CrossRef]
68. Song, Y.; Zhang, S.; Li, J.; Zhao, C.; Zhang, X. Electrodeposition of Ca-P coatings on biodegradable Mg alloy: In vitro biomineralization behavior. *Acta Biomater.* **2010**, *6*, 1736–1742. [CrossRef]
69. Saremi, M.; Mohajernia, S.; Hejazi, S. Controlling the degradation rate of AZ31 Magnesium alloy and purity of nano-hydroxyapatit coating by pulse electrodeposition. *Mater. Lett.* **2014**, *129*, 111–113. [CrossRef]
70. Seyedraoufi, Z.S.; Mirdamadi, S. Effects of pulse electrodeposition parameters and alkali treatment on the properties of nano hydroxyapatite coating on porous MgeZn scaffold for bone tissue engineering application. *Mater. Chem. Phys.* **2014**, *148*, 519–527. [CrossRef]

71. Seyedraoufi, Z.S.; Mirdamadi, S. In vitro biodegradability and biocompatibility of porous Mg-Zn scaffolds coated with nano hydroxyapatite via pulse electrodeposition. *Trans. Nonferrous Met. Soc. China Engl. Ed.* **2015**, *25*, 4018–4027. [CrossRef]
72. Monasterio, N.; Ledesma, J.L.; Aranguiz, I.; Garcia-Romero, A.; Zuza, E. Analysis of electrodeposition processes to obtain calcium phosphate layer on AZ31 alloy. *Surf. Coat. Technol.* **2017**, *319*, 12–22. [CrossRef]
73. Han, J.; Blawert, C.; Tang, S.; Yang, J.; Hu, J.; Zheludkevich, M.L. Effect of surface pre-treatments on the formation and degradation behaviour of a calcium phosphate coating on pure magnesium. *Coatings* **2019**, *9*, 259. [CrossRef]
74. Liao, Y.M.; De Feng, Z.; Li, S.W. Preparation and characterization of hydroxyapatite coatings on human enamel by electrodeposition. *Thin Solid Film.* **2008**, *516*, 6145–6150. [CrossRef]
75. Benea, L.; Danaila, E.; Ponthiaux, P. Effect of titania anodic formation and hydroxyapatite electrodeposition on electrochemical behaviour of Ti-6Al-4V alloy under fretting conditions for biomedical applications. *Corros. Sci.* **2015**, *91*, 262–271. [CrossRef]
76. Liu, S.; Li, H.; Zhang, L.; Yin, X.; Guo, Y. In simulated body fluid performance of polymorphic apatite coatings synthesized by pulsed electrodeposition. *Mater. Sci. Eng. C* **2017**, *79*, 100–107. [CrossRef] [PubMed]
77. Zhao, X.; Chen, X.; Zhang, L.; Liu, Q.; Wang, Y.; Zhang, W.; Zheng, J. Preparation of nano-hydroxyapatite coated carbon nanotube reinforced hydroxyapatite composites. *Coatings* **2018**, *8*, 357. [CrossRef]
78. Li, T.T.; Ling, L.; Lin, M.C.; Jiang, Q.; Lin, Q.; Lou, C.W.; Lin, J.H. Effects of ultrasonic treatment and current density on the properties of hydroxyapatite coating via electrodeposition and its in vitro biomineralization behavior. *Mater. Sci. Eng. C* **2019**, *105*, 110062. [CrossRef] [PubMed]
79. Gopi, D.; Karthika, A.; Nithiya, S.; Kavitha, L. In vitro biological performance of minerals substituted hydroxyapatite coating by pulsed electrodeposition method. *Mater. Chem. Phys.* **2014**, *144*, 75–85. [CrossRef]
80. Drevet, R.; Benhayoune, H. Pulsed electrodeposition for the synthesis of strontium-substituted calcium phosphate coatings with improved dissolution properties. *Mater. Sci. Eng. C* **2013**, *33*, 4260–4265. [CrossRef]
81. Yajing, Y.; Qiongqiong, D.; Yong, H.; Han, S.; Pang, X. Magnesium substituted hydroxyapatite coating on titanium with nanotublar TiO_2 intermediate layer via electrochemical deposition. *Appl. Surf. Sci.* **2014**, *305*, 77–85. [CrossRef]
82. Huang, Y.; Hao, M.; Nian, X.; Qiao, H.; Zhang, X.X.; Zhang, X.X.; Song, G.; Guo, J.; Pang, X.; Zhang, H. Strontium and copper co-substituted hydroxyapatite-based coatings with improved antibacterial activity and cytocompatibility fabricated by electrodeposition. *Ceram. Int.* **2016**, *42*, 11876–11888. [CrossRef]
83. Huang, Y.; Qiao, H.; Nian, X.; Zhang, X.; Zhang, X.; Song, G.; Xu, Z.; Zhang, H.; Han, S. Improving the bioactivity and corrosion resistance properties of electrodeposited hydroxyapatite coating by dual doping of bivalent strontium and manganese ion. *Surf. Coat. Technol.* **2016**, *291*, 205–215. [CrossRef]
84. Fu, D.L.; Jiang, Q.H.; He, F.M.; Fu, B.P. Adhesion of bone marrow mesenchymal stem cells on porous titanium surfaces with strontium-doped hydroxyapatite coating. *J. Zhejiang Univ. Sci. B* **2017**, *18*, 778–788. [CrossRef]
85. Huang, Y.; Wang, W.; Zhang, X.; Liu, X.; Xu, Z.; Han, S.; Su, Z.; Liu, H.; Gao, Y.; Yang, H. A prospective material for orthopedic applications: Ti substrates coated with a composite coating of a titania-nanotubes layer and a silver-manganese-doped hydroxyapatite layer. *Ceram. Int.* **2017**, *44*, 5528–5542. [CrossRef]
86. Karthika, A. Aliovalent ions substituted hydroxyapatite coating on titanium for improved medical applications. *Mater. Today Proc.* **2018**, *5*, 8768–8774. [CrossRef]
87. Drevet, R.; Zhukova, Y.; Dubinskiy, S.; Kazakbiev, A.; Naumenko, V.; Abakumov, M.; Fauré, J.; Benhayoune, H.; Prokoshkin, S. Electrodeposition of cobalt-substituted calcium phosphate coatings on Ti22Nb6Zr alloy for bone implant applications. *J. Alloys Compd.* **2019**, *793*, 576–582. [CrossRef]
88. Furko, M.; Della Bella, E.; Fini, M.; Balázsi, C. Corrosion and biocompatibility examination of multi-element modified calcium phosphate bioceramic layers. *Mater. Sci. Eng. C* **2019**, *95*, 381–388. [CrossRef]
89. Meng, E.C.; Guan, S.K.; Wang, H.X.; Wang, L.G.; Zhu, S.J.; Hu, J.H.; Ren, C.X.; Gao, J.H.; Feng, Y.S. Effect of electrodeposition modes on surface characteristics and corrosion properties of fluorine-doped hydroxyapatite coatings on Mg-Zn-Ca alloy. *Appl. Surf. Sci.* **2011**, *257*, 4811–4816. [CrossRef]
90. Qiu, X.; Wan, P.; Tan, L.; Fan, X.; Yang, K. Preliminary research on a novel bioactive silicon doped calcium phosphate coating on AZ31 magnesium alloy via electrodeposition. *Mater. Sci. Eng. C* **2014**, *36*, 65–76. [CrossRef]
91. Aboudzadeh, N.; Dehghanian, C.; Shokrgozar, M.A. Effect of electrodeposition parameters and substrate on morphology of Si-HA coating. *Surf. Coat. Technol.* **2019**, *375*, 341–351. [CrossRef]

92. Liu, S.J.; Li, H.J.; Zhang, L.L.; Feng, L.; Yao, P. Strontium and magnesium substituted dicalcium phosphate dehydrate coating for carbon/carbon composites prepared by pulsed electrodeposition. *Appl. Surf. Sci.* **2015**, *359*, 288–292. [CrossRef]
93. Peñaflor Galindo, T.G.; Kataoka, T.; Fujii, S.; Okuda, M.; Tagaya, M. Preparation of nanocrystalline zinc-substituted hydroxyapatite films and their biological properties. *Colloids Interface Sci. Commun.* **2016**, *10–11*, 15–19. [CrossRef]
94. Huang, C.N.; Tang, Z.T.; Chan, F.E.; Burnouf, T.; Huang, W.C.; Chen, P.C. Fabrication of co-electrodeposition of plasma proteins/iridium oxide hybrid films. *Ceram. Int.* **2018**, *44*, 117–120. [CrossRef]
95. Gao, X.; Gui, R.; Guo, H.; Wang, Z.; Liu, Q. Creatinine-induced specific signal responses and enzymeless ratiometric electrochemical detection based on copper nanoparticles electrodeposited on reduced graphene oxide-based hybrids. *Sens. Actuators B Chem.* **2019**, *285*, 201–208. [CrossRef]
96. Benea, L.; Celis, J.P. Reactivity of porous titanium oxide film and chitosan layer electrochemically formed on Ti-6Al-4V alloy in biological solution. *Surf. Coat. Technol.* **2018**, *354*, 145–152. [CrossRef]
97. Nakano, K.; Egashira, K.; Masuda, S.; Funakoshi, K.; Zhao, G.; Kimura, S.; Matoba, T.; Sueishi, K.; Endo, Y.; Kawashima, Y.; et al. Formulation of nanoparticle-eluting stents by a cationic electrodeposition coating technology. efficient nano-drug delivery via bioabsorbable polymeric nanoparticle-eluting stents in porcine coronary arteries. *JACC Cardiovasc. Interv.* **2009**, *2*, 277–283. [CrossRef]
98. Jiang, P.L.; Hou, R.Q.; Chen, C.D.; Sun, L.; Dong, S.G.; Pan, J.S.; Lin, C.J. Controllable degradation of medical magnesium by electrodeposited composite films of mussel adhesive protein (Mefp-1) and chitosan. *J. Colloid Interface Sci.* **2016**, *478*, 246–255. [CrossRef]
99. Meng, L.; Li, Y.; Pan, K.; Zhu, Y.; Wei, W.; Li, X.; Liu, X. Colloidal particle based electrodeposition coatings on NiTi alloy: Reduced releasing of nickel ions and improved biocompatibility. *Mater. Lett.* **2018**, *230*, 228–231. [CrossRef]
100. Sak, A.; Moskalewicz, T.; Zimowski, S.; Cieniek, Ł.; Dubiel, B.; Radziszewska, A.; Kot, M.; Łukaszczyk, A. Influence of polyetheretherketone coatings on the Ti-13Nb-13Zr titanium alloy's bio-tribological properties and corrosion resistance. *Mater. Sci. Eng. C* **2016**, *63*, 52–61. [CrossRef]
101. Chakraborty, R.; Seesala, V.S.; Sengupta, S.; Dhara, S.; Saha, P.; Das, K.; Das, S. Comparison of osteoconduction, cytocompatibility and corrosion protection performance of hydroxyapatite-calcium hydrogen phosphate composite coating synthesized in-situ through pulsed electro-deposition with varying amount of phase and crystallinity. *Surf. Interfaces* **2018**, *10*, 1–10. [CrossRef]
102. Shojaee, P.; Afshar, A. Effects of zirconia content on characteristics and corrosion behavior of hydroxyapatite/ZrO2 biocomposite coatings codeposited by electrodeposition. *Surf. Coat. Technol.* **2015**, *262*, 166–172. [CrossRef]
103. Chakraborty, R.; Seesala, V.S.; Sen, M.; Sengupta, S.; Dhara, S.; Saha, P.; Das, K.; Das, S. MWCNT reinforced bone like calcium phosphate—Hydroxyapatite composite coating developed through pulsed electrodeposition with varying amount of apatite phase and crystallinity to promote superior osteoconduction, cytocompatibility and corrosion protection. *Surf. Coat. Technol.* **2017**, *325*, 496–514. [CrossRef]
104. Huang, Y.; Han, S.; Pang, X.; Ding, Q.; Yan, Y. Electrodeposition of porous hydroxyapatite/calcium silicate composite coating on titanium for biomedical applications. *Appl. Surf. Sci.* **2013**, *271*, 299–302. [CrossRef]
105. Huang, Y.; Zhang, H.; Qiao, H.; Nian, X.; Zhang, X.; Wang, W.; Zhang, X.; Chang, X.; Han, S.; Pang, X. Anticorrosive effects and in vitro cytocompatibility of calcium silicate/zinc-doped hydroxyapatite composite coatings on titanium. *Appl. Surf. Sci.* **2015**, *357*, 1776–1784. [CrossRef]
106. Mostafa, N.Y.; Montaser, A.; Al-Affray, R.A.; Kamel, M.M.; Alhadhrami, A. Processing and characterization of novel calcium titanate/Na-titanate nanotube/rutile nanocomposite coating on titanium metal. *Appl. Phys. A Mater. Sci. Process.* **2019**, *125*, 1–6. [CrossRef]
107. Chozhanathmisra, M.; Murugan, N.; Karthikeyan, P.; Sathishkumar, S.; Anbarasu, G.; Rajavel, R. Development of antibacterial activity and corrosion resistance properties of electrodeposition of mineralized hydroxyapatite coated on titanium alloy for biomedical applications. *Mater. Today Proc.* **2017**, *4*, 12393–12400. [CrossRef]
108. Chozhanathmisra, M.; Ramya, S.; Kavitha, L.; Gopi, D. Development of zinc-halloysite nanotube/minerals substituted hydroxyapatite bilayer coatings on titanium alloy for orthopedic applications. *Colloids Surf. A Physicochem. Eng. Asp.* **2016**, *511*, 357–365. [CrossRef]
109. Fathyunes, L.; Khalil-Allafi, J. Characterization and corrosion behavior of graphene oxide-hydroxyapatite composite coating applied by ultrasound-assisted pulse electrodeposition. *Ceram. Int.* **2017**, *43*, 13885–13894. [CrossRef]

110. Fathyunes, L.; Khalil-Allafi, J.; Sheykholeslami, S.O.R.; Moosavifar, M. Biocompatibility assessment of graphene oxide-hydroxyapatite coating applied acon TiO_2 nanotubes by ultrasound-assisted pulse electrodeposition. *Mater. Sci. Eng. C* **2018**, *87*, 10–21. [CrossRef]
111. Zhang, L.; Zhu, F.; Li, H.; Zhao, F.; Li, S. A duplex coating composed of electrophoretic deposited graphene oxide inner-layer and electrodeposited graphene oxide/Mg substituted hydroxyapatite outer-layer on carbon/carbon composites for biomedical application. *Ceram. Int.* **2018**, *44*, 21229–21237. [CrossRef]
112. Gopi, D.; Shinyjoy, E.; Sekar, M.; Surendiran, M.; Kavitha, L.; Sampath Kumar, T.S. Development of carbon nanotubes reinforced hydroxyapatite composite coatings on titanium by electrodeposition method. *Corros. Sci.* **2013**, *73*, 321–330. [CrossRef]
113. Fathyunes, L.; Khalil-Allafi, J.; Moosavifar, M. Development of graphene oxide/calcium phosphate coating by pulse electrodeposition on anodized titanium: Biocorrosion and mechanical behaavior. *J. Mech. Behav. Biomed. Mater.* **2019**, *90*, 575–586. [CrossRef] [PubMed]
114. Khazeni, D.; Saremi, M.; Soltani, R. Development of HA-CNTs composite coating on AZ31 magnesium alloy by cathodic electrodeposition. Part 1: Microstructural and mechanical characterization. *Ceram. Int.* **2019**, *45*, 11174–11185. [CrossRef]
115. Chakraborty, R.; Manna, J.S.; Das, D.; Sen, M.; Saha, P. A comparative outlook of corrosion behaviour and chlorophyll assisted growth kinetics of various carbon nano-structure reinforced hydroxyapatite-calcium orthophosphate coating synthesized in-situ through pulsed electrochemical deposition. *Appl. Surf. Sci.* **2019**, *475*, 28–42. [CrossRef]
116. Długoń, E.; Niemiec, W.; Frączek-Szczypta, A.; Jeleń, P.; Sitarz, M.; Błazewicz, M. Spectroscopic studies of electrophoretically deposited hybrid HAp/CNT coatings on titanium. *Spectrochim. Acta Part. A Mol. Biomol. Spectrosc.* **2014**, *133*, 872–875. [CrossRef]
117. Maleki-Ghaleh, H.; Khalil-Allafi, J. Characterization, mechanical and in vitro biological behavior of hydroxyapatite-titanium-carbon nanotube composite coatings deposited on NiTi alloy by electrophoretic deposition. *Surf. Coat. Technol.* **2019**, *363*, 179–190. [CrossRef]
118. Poorraeisi, M.; Afshar, A. The study of electrodeposition of hydroxyapatite-ZrO_2-TiO_2 nanocomposite coatings on 316 stainless steel. *Surf. Coat. Technol.* **2018**, *339*, 199–207. [CrossRef]
119. Pei, L.; Zhang, B.; Luo, H.; Wu, X.; Li, G.; Sheng, H.; Zhang, L. Electrodeposition of ZnO Nanoprism-Zn substituted hydroxyapatite duplex layer coating for carbon fiber. *Ceram. Int.* **2019**, *45*, 14278–14286. [CrossRef]
120. Göncü, Y.; Geçgin, M.; Bakan, F.; Ay, N. Electrophoretic deposition of hydroxyapatite-hexagonal boron nitride composite coatings on Ti substrate. *Mater. Sci. Eng. C* **2017**, *79*, 343–353. [CrossRef]
121. Khanmohammadi, S.; Ilkhchi, M.O.; Khalil-Allafi, J. Electrophoretic deposition and characterization of bioglass-whisker hydroxyapatite nanocomposite coatings on titanium substrate. *Surf. Coat. Technol.* **2019**, *378*, 124949. [CrossRef]
122. Yılmaz, E.; Çakıroğlu, B.; Gökçe, A.; Findik, F.; Gulsoy, H.O.; Gulsoy, N.; Mutlu, Ö.; Özacar, M. Novel hydroxyapatite/graphene oxide/collagen bioactive composite coating on Ti16Nb alloys by electrodeposition. *Mater. Sci. Eng. C* **2019**, *101*, 292–305. [CrossRef]
123. Kumar, A.M.; Adesina, A.Y.; Hussein, M.A.; Ramakrishna, S.; Al-Aqeeli, N.; Akhtar, S.; Saravanan, S. PEDOT/FHA nanocomposite coatings on newly developed Ti-Nb-Zr implants: Biocompatibility and surface protection against corrosion and bacterial infections. *Mater. Sci. Eng. C* **2019**, *98*, 482–495. [CrossRef] [PubMed]
124. Chakraborty, R.; Seesala, V.S.; Manna, J.S.; Saha, P.; Dhara, S. Synthesis, characterization and cytocompatibility assessment of hydroxyapatite-polypyrrole composite coating synthesized through pulsed reverse electrochemical deposition. *Mater. Sci. Eng. C* **2019**, *94*, 597–607. [CrossRef] [PubMed]
125. Pang, X.; Zhitomirsky, I. Electrodeposition of hydroxyapatite-silver-chitosan nanocomposite coatings. *Surf. Coat. Technol.* **2008**, *202*, 3815–3821. [CrossRef]
126. Yan, Y.; Zhang, X.; Li, C.; Huang, Y.; Ding, Q.; Pang, X. Preparation and characterization of chitosan-silver/hydroxyapatite composite coatings onTiO_2 nanotube for biomedical applications. *Appl. Surf. Sci.* **2015**, *332*, 62–69. [CrossRef]
127. Zhao, P.; Zhao, Y.; Xiao, L.; Deng, H.; Du, Y.; Chen, Y.; Shi, X. Electrodeposition to construct free-standing chitosan/layered double hydroxides hydro-membrane for electrically triggered protein release. *Colloids Surf. B Biointerfaces* **2017**, *158*, 474–479. [CrossRef]

128. Zhong, Z.; Qin, J.; Ma, J. Electrophoretic deposition of biomimetic zinc substituted hydroxyapatite coatings with chitosan and carbon nanotubes on titanium. *Ceram. Int.* **2015**, *41*, 8878–8884. [CrossRef]
129. Clifford, A.; Lee, B.E.J.; Grandfield, K.; Zhitomirsky, I. Biomimetic modification of poly-L-lysine and electrodeposition of nanocomposite coatings for orthopaedic applications. *Colloids Surf. B Biointerfaces* **2019**, *176*, 115–121. [CrossRef]
130. Baştan, F.E.; Atiq Ur Rehman, M.; Avcu, Y.Y.; Avcu, E.; Üstel, F.; Boccaccini, A.R. Electrophoretic co-deposition of PEEK-hydroxyapatite composite coatings for biomedical applications. *Colloids Surf. B Biointerfaces* **2018**, *169*, 176–182. [CrossRef]
131. Patiño-Herrera, R.; González-Alatorre, G.; Estrada-Baltazar, A.; Escoto-Chavéz, S.E.; Pérez, E. Hydrophobic coatings for prevention of dental enamel erosion. *Surf. Coat. Technol.* **2015**, *275*, 148–154. [CrossRef]
132. Khabazian, S.; Sanjabi, S. The effect of multi-walled carbon nanotube pretreatments on the electrodeposition of Ni-MWCNTs coatings. *Appl. Surf. Sci.* **2011**, *257*, 5850–5856. [CrossRef]
133. El-Wassefy, N.A.; Reicha, F.M.; Aref, N.S. Electro-chemical deposition of nano hydroxyapatite-zinc coating on titanium metal substrate. *Int. J. Implant. Dent.* **2017**, *3*, 1–8. [CrossRef] [PubMed]
134. Fu, C.; Zhang, X.; Savino, K.; Gabrys, P.; Gao, Y.; Chaimayo, W.; Miller, B.L.; Yates, M.Z. Antimicrobial silver-hydroxyapatite composite coatings through two-stage electrochemical synthesis. *Surf. Coat. Technol.* **2016**, *301*, 13–19. [CrossRef]
135. Bartmanski, M.; Cieslik, B.; Glodowska, J.; Kalka, P.; Pawlowski, L.; Pieper, M.; Zielinski, A. Electrophoretic deposition (EPD) of nanohydroxyapatite-nanosilver coatings on Ti13Zr13Nb alloy. *Ceram. Int.* **2017**, *43*, 11820–11829. [CrossRef]
136. Bartmanski, M. The properties of nanosilver—Doped nanohydroxyapatite coating on the Ti13Zr13Nb alloy. *Adv. Mater. Sci.* **2017**, *17*, 18–28. [CrossRef]
137. Bartmanski, M.; Zielinski, A.; Jazdzewska, M.; Głodowska, J.; Kalka, P. Effects of electrophoretic deposition times and nanotubular oxide surfaces on properties of the nanohydroxyapatite/nanocopper coating on the Ti13Zr13Nb alloy. *Ceram. Int.* **2019**, *45*, 20002–20010. [CrossRef]
138. Majkowska-Marzec, B.; Rogala-Wielgus, D.; Bartmański, M.; Bartosewicz, B.; Zieliński, A.S. Comparison of Properties of the Hybrid and Bilayer MWCNTs—Hydroxyapatite Coatings on Ti Alloy. *Coatings* **2019**, *9*, 643. [CrossRef]
139. Jang, J.M.; Kim, S.D.; Park, T.E.; Choe, H.C. Ultra-fine structures of Pd-Ag-HAp nanoparticle deposition on protruded TiO$_2$ barrier layer for dental implant. *Appl. Surf. Sci.* **2018**, *432*, 285–293. [CrossRef]
140. Silva-Ichante, M.; Reyes-Vidal, Y.; Bácame-Valenzuela, F.J.; Ballesteros, J.C.; Arciga, E.; Ţălu, Ş.; Méndez-Albores, A.; Trejo, G. Electrodeposition of antibacterial Zn-Cu/silver nanoparticle (AgNP) composite coatings from an alkaline solution containing glycine and AgNPs. *J. Electroanal. Chem.* **2018**, *823*, 328–334. [CrossRef]
141. Kumar, A.M.; Nagarajan, S.; Ramakrishna, S.; Sudhagar, P.; Kang, Y.S.; Kim, H.; Gasem, Z.M.; Rajendran, N. Electrochemical and in vitro bioactivity of polypyrrole/ceramic nanocomposite coatings on 316L SS bio-implants. *Mater. Sci. Eng. C* **2014**, *43*, 76–85. [CrossRef]
142. Lázaro-Martínez, J.M.; Byrne, A.J.; Rodríguez-Castellón, E.; Manrique, J.M.; Jones, L.R.; Dall'Orto, V.C. Linear polyethylenimine-decorated gold nanoparticles: One-step electrodeposition and studies of interaction with viral and animal proteins. *Electrochim. Acta* **2019**, *301*, 126–135. [CrossRef]
143. Wang, Z.; Zhang, X.; Gu, J.; Yang, H.; Nie, J.; Ma, G. Electrodeposition of alginate/chitosan layer-by-layer composite coatings on titanium substrates. *Carbohydr. Polym.* **2014**, *103*, 38–45. [CrossRef] [PubMed]
144. Kumar, A.M.; Suresh, B.; Das, S.; Obot, I.B.; Adesina, A.Y.; Ramakrishna, S. Promising bio-composites of polypyrrole and chitosan: Surface protective and in vitro biocompatibility performance on 316L SS implants. *Carbohydr. Polym.* **2017**, *173*, 121–130. [CrossRef] [PubMed]
145. Pawłowski, Ł.; Bartmański, M.; Strugała, G.; Mielewczyk-Gryń, A.; Jażdżewska, M.; Zieliński, A. Electrophoretic deposition and characterization of Chitosan/Eudragit E 100 coatings on titanium substrate. *Coatings* **2020**, *10*, 607. [CrossRef]
146. Bonifacio, M.A.; Cometa, S.; Dicarlo, M.; Baruzzi, F.; de Candia, S.; Gloria, A.; Giangregorio, M.M.; Mattioli-Belmonte, M.; De Giglio, E. Gallium-modified chitosan/poly(acrylic acid) bilayer coatings for improved titanium implant performances. *Carbohydr. Polym.* **2017**, *166*, 348–357. [CrossRef]

147. Trzaskowska, P.A.; Poniatowska, A.; Tokarska, K.; Wiśniewski, C.; Ciach, T.; Malinowska, E. Promising electrodeposited biocompatible coatings for steel obtained from polymerized microemulsions. *Colloids Surf. A Physicochem. Eng. Asp.* **2020**, *591*, 124555. [CrossRef]
148. Liu, L.; Qi, W.; Gao, X.; Wang, C.; Wang, G. Synergistic effect of metal ion additives on graphitic carbon nitride nanosheet-templated electrodeposition of Cu@CuO for enzyme-free glucose detection. *J. Alloys Compd.* **2018**, *745*, 155–163. [CrossRef]
149. Zhao, Z.W.; Zhang, G.; Li, H.G. Preparation of calcium phosphate coating on pure titanium substrate by electrodeposition method. *J. Cent. South. Univ. Technol. Engl. Ed.* **2004**, *11*, 147–151. [CrossRef]
150. Hou, X.; Liu, X.; Xu, J.; Shen, J.; Liu, X. A self-optimizing electrodeposition process for fabrication of calcium phosphate coatings. *Mater. Lett.* **2001**, *50*, 103–107. [CrossRef]
151. Manso, M.; Jiménez, C.; Morant, C.; Herrero, P.; Martínez-Duart, J. Electrodeposition of hydroxyapatite coatings in basic conditions. *Biomaterials* **2000**, *21*, 1755–1761. [CrossRef]
152. Beaufils, S.; Rouillon, T.; Millet, P.; Le Bideau, J.; Weiss, P.; Chopart, J.P.; Daltin, A.L. Synthesis of calcium-deficient hydroxyapatite nanowires and nanotubes performed by template-assisted electrodeposition. *Mater. Sci. Eng. C* **2019**, *98*, 333–346. [CrossRef]
153. Khalili, V.; Khalil-Allafi, J.; Frenzel, J.; Eggeler, G. Bioactivity and electrochemical behavior of hydroxyapatite-silicon-multi walled carbon nano-tubes composite coatings synthesized by EPD on NiTi alloys in simulated body fluid. *Mater. Sci. Eng. C* **2017**, *71*, 473–482. [CrossRef] [PubMed]
154. Pawlik, A.; Rehman, M.A.U.; Nawaz, Q.; Bastan, F.E.; Sulka, G.D.; Boccaccini, A.R. Fabrication and characterization of electrophoretically deposited chitosan-hydroxyapatite composite coatings on anodic titanium dioxide layers. *Electrochim. Acta* **2019**, *307*, 465–473. [CrossRef]
155. Molaei, A.; Yari, M.; Afshar, M.R. Modification of electrophoretic deposition of chitosan-bioactive glass-hydroxyapatite nanocomposite coatings for orthopedic applications by changing voltage and deposition time. *Ceram. Int.* **2015**, *41*, 14537–14544. [CrossRef]
156. Durdu, S.; Usta, M.; Berkem, A.S. Bioactive coatings on Ti6Al4V alloy formed by plasma electrolytic oxidation. *Surf. Coat. Technol.* **2016**, *301*, 85–93. [CrossRef]
157. Liu, S.; Li, B.; Liang, C.; Wang, H.; Qiao, Z. Formation mechanism and adhesive strength of a hydroxyapatite/TiO_2 composite coating on a titanium surface prepared by micro-arc oxidation. *Appl. Surf. Sci.* **2016**, *362*, 109–114. [CrossRef]
158. Shi, M.; Li, H. The morphology, structure and composition of microarc oxidation (MAO) ceramic coating in Ca-P electrolyte with complexing agent EDTMPS and interpretation hypothesis of MAO process. *Surf. Eng. Appl. Electrochem.* **2016**, *52*, 32–42. [CrossRef]
159. Mao, Y.; Yan, J.; Wang, L.; Dong, W.; Jia, Y.; Hu, X.; Wang, X. Formation and properties of bioactive barium titanate coatings produced by plasma electrolytic oxidation. *Ceram. Int.* **2018**, *44*, 12978–12986. [CrossRef]
160. Luo, S.; Wang, Q.; Ye, R.; Ramachandran, C.S. Effects of electrolyte concentration on the microstructure and properties of plasma electrolytic oxidation coatings on Ti-6Al-4V alloy. *Surf. Coat. Technol.* **2019**, *375*, 864–876. [CrossRef]
161. Zhang, X.; Yu, Y.; Jiang, D.; Jiao, Y.; Wu, Y.; Peng, Z.; Zhou, J.; Wu, J.; Dong, Z. Synthesis and characterization of a bi-functional hydroxyapatite/Cu-doped TiO_2 composite coating. *Ceram. Int.* **2019**, *45*, 6693–6701. [CrossRef]
162. Dziaduszewska, M.; Shimabukuro, M.; Seramak, T.; Zielinski, A. Effects of micro-arc oxidation process parameters on characteristics of calcium-phosphate containing oxide layers on the selective laser melted Ti13Zr13Nb alloy. *Coatings* **2020**, *10*, 745. [CrossRef]
163. Antonio, R.F.; Rangel, E.C.; Mas, B.A.; Duek, E.A.R.; Cruz, N.C. Growth of hydroxyapatite coatings on tantalum by plasma electrolytic oxidation in a single step. *Surf. Coat. Technol.* **2019**, *357*, 698–705. [CrossRef]
164. Cengiz, S.; Azakli, Y.; Tarakci, M.; Stanciu, L.; Gencel, Y. Microarc oxidation type discharge types and bio properties of the coating synthesized on zirconium. *Mater. Sci. Eng. C* **2017**, *77*, 374–383. [CrossRef] [PubMed]
165. Kim, Y.K.; Jang, Y.S.; Kim, S.Y.; Lee, M.H. Functions achieved by the hyaluronic acid derivatives coating and hydroxide film on bio-absorbed Mg. *Appl. Surf. Sci.* **2019**, *473*, 31–39. [CrossRef]
166. Durdu, S.; Aktuğ, S.L.; Korkmaz, K. Characterization and mechanical properties of the duplex coatings produced on steel by electro-spark deposition and micro-arc oxidation. *Surf. Coat. Technol.* **2013**, *236*, 303–308. [CrossRef]

167. Durdu, S.; Korkmaz, K.; Aktuğ, S.L.; Çakır, A. Characterization and bioactivity of hydroxyapatite-based coatings formed on steel by electro-spark deposition and micro-arc oxidation. *Surf. Coat. Technol.* **2017**, *326*, 111–120. [CrossRef]
168. Aliyu, A.A.; Abdul-Rani, A.M.; Rao, T.V.V.L.N.; Axinte, E.; Hastuty, S.; Parameswari, R.P.; Subramaniam, J.R.; Thyagarajan, S.P. Characterization, adhesion strength and in-vitro cytotoxicity investigation of hydroxyapatite coating synthesized on Zr-based BMG by electro discharge process. *Surf. Coat. Technol.* **2019**, *370*, 213–226. [CrossRef]
169. Devgan, S.; Sidhu, S.S. Surface modification of β-type titanium with multi-walled CNTs/ μ-HAp powder mixed electro discharge treatment process. *Mater. Chem. Phys.* **2020**, *239*, 122005. [CrossRef]
170. Momeni, M.M.; Nazari, Z. Pt/PANI-MWCNTs nanocomposite coating prepared by electropolymerisation-electrodeposition for glycerol electro-oxidation. *Surf. Eng.* **2015**, *31*, 472–479. [CrossRef]
171. Rikhari, B.; Mani, S.P.; Rajendran, N. Electrochemical behavior of polypyrrole/chitosan composite coating on Ti metal for biomedical applications. *Carbohydr. Polym.* **2018**, *189*, 126–137. [CrossRef]
172. Almeida, L.C.; Correia, R.D.; Squillaci, G.; Morana, A.; La Cara, F.; Correia, J.P.; Viana, A.S. Electrochemical deposition of bio-inspired laccase-polydopamine films for phenolic sensors. *Electrochim. Acta* **2019**, *319*, 462–471. [CrossRef]

© 2020 by the authors. Licensee MDPI, Basel, Switzerland. This article is an open access article distributed under the terms and conditions of the Creative Commons Attribution (CC BY) license (http://creativecommons.org/licenses/by/4.0/).

MDPI\
St. Alban-Anlage 66\
4052 Basel\
Switzerland\
Tel. +41 61 683 77 34\
Fax +41 61 302 89 18\
www.mdpi.com

Coatings Editorial Office\
E-mail: coatings@mdpi.com\
www.mdpi.com/journal/coatings

www.ingramcontent.com/pod-product-compliance
Lightning Source LLC
LaVergne TN
LVHW070644100526
838202LV00013B/873